Physiology of Finfish and Shellfish

(Strictly as per Syllabus for B.F.Sc. and M.F.Sc. Students)

NIPA® GENX ELECTRONIC RESOURCES & SOLUTIONS P. LTD.
New Delhi-110 034

Physiology of Finfish and Shellfish

(Strictly as per Syllabus for B.F.Sc. and M.F.Sc. Students)

Prof. (Dr.) Kasturi Samantaray
College of Fisheries
Orissa University of Agriculture and Technology
Rangeilunda, Berhampur – 760 007
Ganjam, Odisha

NIPA® GENX ELECTRONIC RESOURCES & SOLUTIONS P. LTD.
New Delhi-110 034

NIPA® GENX ELECTRONIC
RESOURCES & SOLUTIONS P. LTD.

101,103, Vikas Surya Plaza, CU Block
L.S.C. Market, Pitam Pura, New Delhi-110 034
Ph : +91 11 27341616, 27341717, 27341718
E-mail: newindiapublishingagency@gmail.com
web: www.nipabooks.com

For customer assistance, please contact
Phone: + 91-11-27 34 17 17
Fax: + 91-11- 27 34 16 16
E-Mail: feedbacks@nipabooks.com

ISBN: 978-81-19254-00-2

Composed and Designed by NIPA®.

Dedicated to

My Parents
Smt. Sabita Samantaray
Sri Kishore Chandra Samantaray,
and
My Professor
Prof. Asit Baran Das

डा. एस. अय्यप्पन
सचिव एवं महानिदेशक
Dr. S. AYYAPPAN
SECRETARY & DIRECTOR GENERAL

भारत सरकार
कृषि अनुसंधान और शिक्षा विभाग एवं
भारतीय कृषि अनुसंधान परिषद
कृषि मंत्रालय, कृषि भवन, नई दिल्ली 110 001

GOVERNMENT OF INDIA
DEPARTMENT OF AGRICULTURAL RESEARCH OF EDUCATION
AND
INDIAN COUNCIL OF AGRICULTURAL RESERACH
MINISTRY OF AGRICULTURE, KRISHI BHAVAN, NEW DELHI 110 001
Tel.: 23382629; 23386711 Fax: 91-11-23384773
E-mail: dg.icer@nic.in

Foreword

I am happy to articulate this book titled "Physiology of Finfish and Shellfish" authored by Prof. Kasturi Samantaray, College of Fisheries, Odisha University of Agriculture and Technology, Rangeilunda, Berhampur, which is one among a few textbooks available in the discipline of Fishery Science that covers important physiological aspects involved in the process of digestion, respiration, circulation, nervous, osmoregulation, reproduction etc., pertaining to both finfishes and shellfishes (decapods including prawn and crab).

India is a carp country and carps are the mainstay of inland freshwater production of India. Hence the chapter particularly physiology of reproduction of this book shall help fishery scientist and farmers to find out new breeding strategies to augment the gap of seed production which is the unique critical input for successful culture. Furthermore, the author has given tremendous importance to the learners in the fishery discipline and designed the book looking into the basic needs of the students. This book strictly adheres to the syllabi implemented for the B. F. Sc. and M. F. Sc. students of Agricultural Universities and allied sectors.

I hope the book would certainly help the students, researchers and farmers working in the field of fishery science for the development of new strategies to improve the fishery sector in the recent era.

I wish her a great success.

Dated the 23rd May, 2014
New Delhi

(Dr. S. Ayyappan)
Director General, ICAR &
Secretary, DARE

Preface

World aquaculture is increasing day-by-day from its infancy of layman's culture to scientific culture, for the enhancement of production to meet the growing demand of the human being through optimum utilization of available water resources. This is possible due to the scientific knowledge on the fish and shellfish. Physiology is the scientific study on the functioning of the various systems of the organism which compliments with the anatomy. The present work is a handy book on fish and shellfish (decapod) physiology. The author also gives an informative prospective on the basic anatomical structures and functions of important systems of both finfish and shellfish.

The author found that many a time the aqua-culturists had great difficulties in solving the common problems encountered during the culture. This is because of the fact that they lack basic knowledge on the physiology of cultured species. By providing the primary information on physiology through this book, the fundamental problems can be partly addressed and their culture can be efficiently managed.

This book on Physiology of Finfish and Shellfish has been written after the author's 25 years of lecturer on the subject both at undergraduate and postgraduate level. Since this book is based on the prescribed syllabus on the subject, it will be useful for the students of Fishery Science. Many books are available on the Physiology of either finfish or shellfish. However not a single book is available in the market till now dealing both finfish and shellfish physiology. Therefore, the author had made an attempt to bridge the gap. Whatever physiology books are available they are very elaborative for the research or expensive only for the libraries. Thus, these are not suitable either for the ready reference or budget of the common aqua culturists and beginners of Fishery Science. This will be an informative handy book which will be easily referred and also affordable. So, this book will be used by a broad spectrum of fish and shellfish students, biologists, research workers and progressive aqua-culturists.

Prof. Kasturi Samantaray

Acknowledgement

Inspirations and encouragements of my brothers (Falguni, Kiran, Pradeepta, Kiriti and Kaushik), Sisters (Kalyani, Kaumudi, Basanti, Basantamanjari and Susipta) and my children (Kunal, Kritarth, Kumuda and Krutikesh) are gratefully acknowledged.

Contents

CHAPTER - 1

Finfish and Shellfish

Aquaculture is the culture of aquatic organisms per unit water area per unit time for maximum profit. The economically important finfishes and shellfishes, whose body physiology is adjustable to varied environmental conditions, are considered as aquaculture candidate species. So, an overall knowledge on the physiology of such species is required to improve the culture practices.

1.1. FINFISH

A fish may be defined as a vertebrate, adapted for a purely aquatic life, propelling and balancing itself by means of fins, and obtaining oxygen (O_2) from the water for survival and life processes. As the fishes have fins, they are also known as finfishes. Fish describes a life from rather than a taxonomic group. They share certain features with other vertebrates. These are gill slits, a notochord, a dorsal hollow nerve cord and tail. In number of individuals, and may be also in number of species, fishes are presently superior to other vertebrates, including mammals. There are 20,000 different species of fishes. Fishes form largest group of vertebrates, Pisces. The various types of fishes differ so much in shape, color and size that it is difficult to believe they all belong to the same group of animals. Fish in general are cold-blooded and adjusted either to freshwater, brackishwater or marine environment.

The fishes have some common characters. These are :

1. Notochord is partially replaced by cartilaginous or bony vertebrae.
2. The endoskeleton may be cartilaginous (Chondrichthyes) or bony (Osteichthyes) and thus the fishes are known as

cartilaginous or bony, respectively. The cartilaginous tissues are firm and elastic, and its matrix is formed of protein called chondrion, secreted by chondroblast. In bony fishes, the matrix contains cells called osteocytes.

3. Presence of gill for respiration.
4. Development of fins for swimming and balancing.
5. An exoskeleton of scales is present in most fishes but some species are naked (without scales) secondarily.
6. Paired visceral arches are present. The first pair forms the upper and lower jaws, the second pair forms suspensorium and the rest supports the gills.
7. The middle ear is absent.
8. Kidney is mesonephric.
9. A swim bladder is usually present (except in some elasmobranches) but is secondarily lost in some species.

Adaptation of all organisms to the environment may be of great importance for their survival in the struggle for existence. Peter Arthur Thomson defines it as "An adaptation is a special adjustment of structure and /or function to meet particular conditions of life". But Khanna (1992) says that it is the capability of animals to change themselves accordingly to the change in environment. In the coastal region or in freshwater, many diverse forms of fishes live under the same conditions with quite different shapes and structures.

1.1.1. Shape and Size

In most cases, the body is elongated and greatly shortened compared to other fishes. The body is flattened in some fishes, particularly in bottom dwelling species and laterally compressed in others. The positions of mouth, eyes, nostrils and gill openings vary widely. Fishes have a streamlined body so that their body is naturally buoyant in water medium. Fishes do not have any neck and so the head bends smoothly in to the truck which narrows into tail (Fig- 1.1). The shape varies according to species. For example trout is boat shaped, eel is like a snake, sea-horses look like horse standing on its tail, rays are flattened from top to bottom, angler-fish and stone-fish resembles rocks, pipefish looks like aquatic weeds and so on.

The body shape is modified depending on the habitat and mode of living of the fish. Accordingly, fish may be cartegorized into different types (Khanna, 1992), such as -

1. Torpediform (Fusiform) - The body is like torpedo, which is suitable for fast swimming. Example - Dogfish, Mullets.

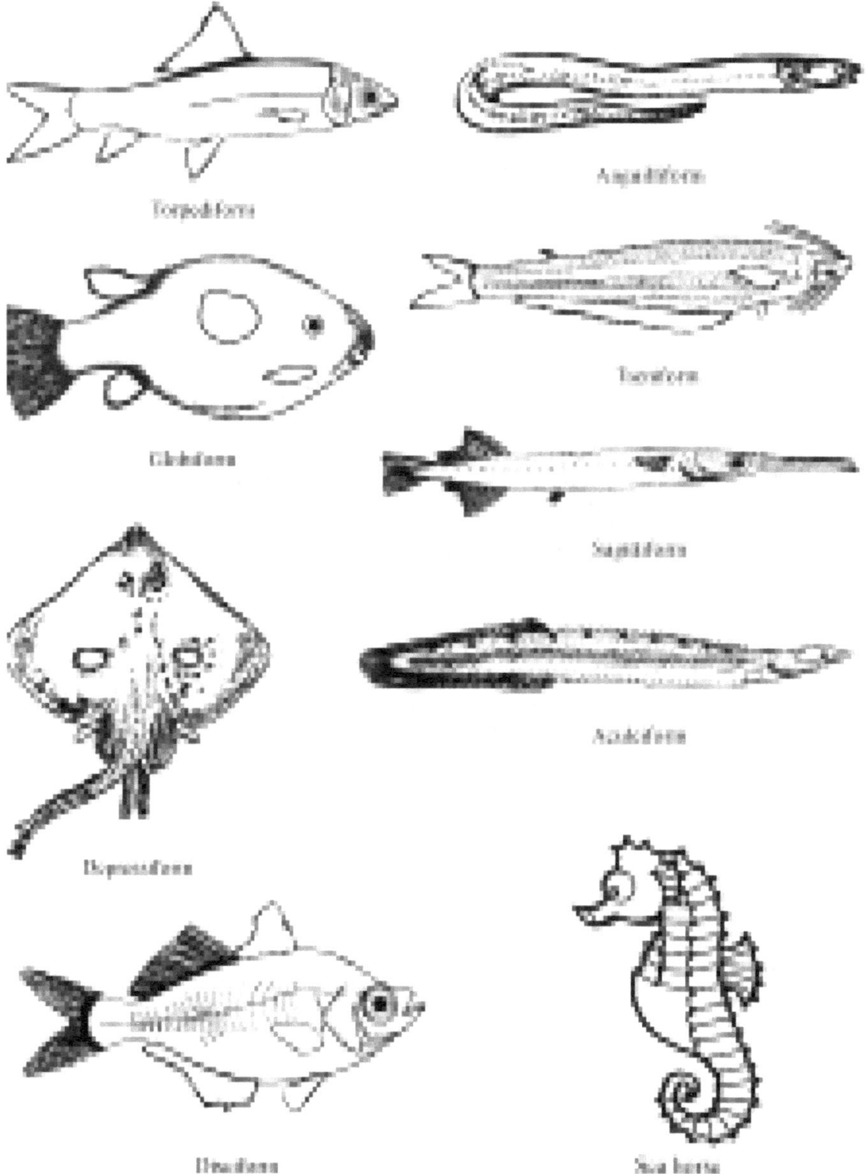

Fig. 1.1 : Variation in the shape of fish.

2. Anguilliform - Body is snake like. Example - *Anguilla, Monopterus*.

3. Globiform - The body is almost rounded or spherical. Example- Puffer fish.
4. Taeniform - Body is elongated, thin and flattened like ribbon. Example- *Trichiurus*.
5. Depressiform- The body is dorso-ventrally flattened. This is the case in deep sea fishes. Example - Skates.
6. Sagittiform - The body is arrow shaped. Example - Pike.
7. Aculeiform - The body is needle-like. Example- Pipe fish
8. Disciform - Body is flattened from side to side. Example- Flat fish.

The size of the fish varies according to species. The smallest fish, known as goby, is the smallest vertebrate in the world. It is only 12-14 mm (about half inch) in size and has a name longer than its size, *Mistichthys luzonensis*. Many coral reef fishes are only 3 to 4 inches in length. Largest fishes are sharks and rays found in sea. The largest fish is the whale shark, which may be more than 40 to 50 feet (12 meters) in length and weigh up to 15 metric tons. The eagle-rays are extra ordinarily monsters in size.

1.1.2. Fins

The fins may be elaborately extended, forming intricate shapes or they may be reduced or lost. Fins may be present all around the fish body, from the back of the head, along the back (dorsal fin) to the tail (caudal fin), along the lower edge (anal fin) (Fig- 1.2). Fins are of two types median (or unpaired) and paired fins, supported by soft or spiny rays of dorsal origin.

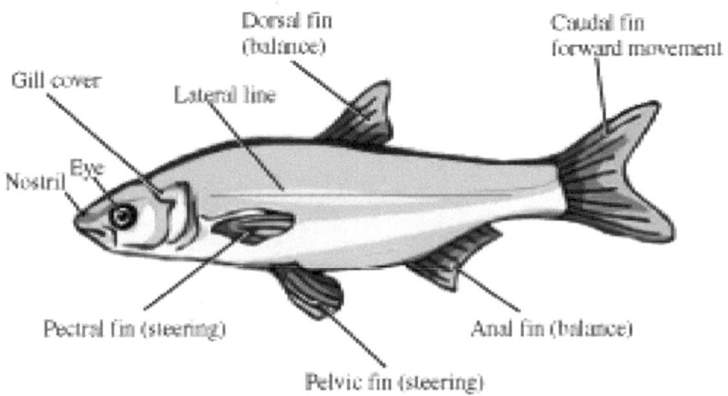

Fig. 1.2 : General morphology of fish.

The median fins include a dorsal fin, anal fin and caudal fin. The dorsal fin lies on the back, the anal fin on the ventral side, behind the vent, and caudal fin at the end of the tail. The median fins develop as a result of differentiation in a continuous embryonic fin fold. During development, a continuous fold of tissue is formed first, running dorsally along the back upto the tip of the tail and is then continued as a ventral fold upto the cloaca. This fold is further strengthened by a series of cartilaginous rods. This is the primitive condition of median fin, which is found in lampreys. But in higher fishes, separate dorsal, caudal and anal fins are formed by the concentration of the radials in particular areas and the degeneration of the fold in the intervening spaces between the fins. The dorsal and anal fins of higher bony fishes are provided with a fleshy muscular lobe at the base surrounding the basals and radials. The dorsal fins are sometimes found in caudal region (*Trichopsis vittatus*) or it may be altogether absent (*Xenomystus*). It may be reduced in size (*Notopterus chitala*) or it may be continuous with caudal fin (*Xiphasia setifer*). Sometimes, there may be two or three dorsal fins (*Gadus morhua*) and the second one may be spiny (*Mugil cephalus*). It may be wing-like in angel fish (*Pterophyllum scalare*), bizarre-shaped (*Pterosis volitans*) or highly extended or modified.

The caudal fin differs from other median fins in its supporting skeleton and it is of three main types.

1. The protocercal (first tail) is the most primitive one. The hind end of the notochord or the vertebral column is straight or divides the caudal fin into two equal lobes, epichordal (dorsal) and hypochordal (ventral) lobes.

2. The heterocercal (unequal) tail is characteristic of chondrichthyes and some primitive bony fishes. In this case the hind end of the notochord is bend upwards and continues almost upto the end of the caudal fin. The hypochordal lobe is much larger than the epichordal, so that the caudal fin is asymmetrical.

3. The homocercal (equal tail) is characteristics of most higher bony fishes. It is symmetrical externally consisting of equal sized epi- and hypo-chordal lobes. But internally the tail is asymmetrical and hinder end of the vertebral column is turned upward and greatly shortened. The end of the vertebral column does not reach the posterior limit of the fin.

The homocercal tail is derived from the heterocercal type. Intermediate types are found in many bony fishes. In Holostei (*Amia* and *Lepidosteus*) the tail shows intermediate condition between heterocercal and homocercal and is called abbreviated heterocercal type. The vertebral column is reduced and withdrawn to the base of the caudal fin, the hypochordal lobe is much enlarged but the epichordal is reduced to a vestige. Adult living dipnoi have symmetrical tail.

Higher teleosts possess externally symmetrical tail. The notochord is much shortened and its upturned region is enclosed in a urostyle which represents the fused center of the posterior caudal vertebrae. In some teleosts the urostyle is fused with caudal hypural and reduced. In many teleosts, the tail is tapering and symmetrical and is called isocercal or leptocercal. The paired fins are the pectoralis and pelvic, corresponding to the fore and hind limbs of the terrestrial vertebrates, respectively. The paired fins are absent in cyclostomes. Cyclostomes are probably the degenerate descendants of Ostracoderms. It appears that the lampreys and hagfishes have probably lost their pair fins, which were possibly present in the Ostracoderms. The bony fish have rayed fins. Of course, some primitive bony fishes have lobed fins, which are less flexible than the rayed fins.

Fins help to maintain the vertical balance when fish is resting or not swimming rapidly. Fins are the main organs for swimming. Sting-ray fish uses its fins to fly rather than to swim. The pectoral fins are employed not only in flying or parachuting, but also for padding, feeling the ground and holding on to stones, keeping the water in motion over the eggs and even to go backward. Many species of fish possess barbles which are the organs for touch.

In sharks, the fins are stiffened at their bases by cartilaginous plates and a series of smaller radial cartilages. Stiff bristle-like collagenous ceratotrichia pass outwards from the radials to the edges of the fin. Such fins are fixed, and cannot alter in shape or be furled, but they can be tilted by muscles attached to the basal cartilages. In rays, separate muscles attach along each elongate jointed radial of the paired fins which are consequently much more mobile, and flex up and down as the ray wings its way. Many rays can walk slowly using the first few radials of the pelvic fin as legs.

1.1.3. Scales

In the evolution of fishes, scales have played an important role. Scales although characteristic feature of majority of fish, they are absent in many fishes, particularly in catfishes. Very primitive fishes usually had thick bony scales in several layers of bone while more evolved fishes have scales of bone, which allow much more freedom of body motion. The scales of fish originate from the mesoderm (skin); they may be similar in structure to teeth.

The structure of scales is also important tool to classify fish. According to the origin of the scales, they are either epidermal as well as dermal derivatives (in elasmobranchs) or exclusively dermal derivative (in teleost). Scales are usually made up of three layers - the outermost thin, enamel like layer called vitrodentine, the middle non-cellular dentine-like layer called cosmine and innermost vascularized bone-like layer called isopedine. Structurally, scales are categorized into five types, such as -

1. **Cosmoid -** These are found in living (*Latimeria*) and extinct (*Crossopterygii*) forms. This type of scales remains deeper in the dermis and consists of three layers. The outer cosmic layer contains a hard glossy, vitrodentine in its outer surface. The middle vascular layer contains numerous anastomosing canals for the blood vessels and the innermost layer is composed of isopedine.
2. **Ganoid (or rhomboid) -** This type of scale mostly found in *Actinopterygii*. The enamel- like outer layer is called ganoine, the middle and inner layers are cosmine and isopedine, respectively. The characteristic feature of this type of scale is that it grows in both dorsal and ventral surface. These scales are formed as simple plates, known as ganoid in the inner layer of the skin. These are roughly four-sided and rhomboid in shape.
3. **Placoid -** This type of scale is found in sharks. Each scale has a disc-like basal plate and trident spine. The spines have vitrodentine, enamel-like outer coating and a inner dentine. A pulp cavity is enclosed within the dentine and the basal plate has an aperture for blood vessels and nerves.
4. **Cycloid -** Teleosts have cycloid scales. These are circular or oval in shape. The free edges of these scales are more or less rounded with no distinct ridges. These ridges are like rings one after another.
5. **Ctenoid -** The spiny rayed teleosts have this type of scale. Structurally they are very much similar with the cycloid scales with only one difference that is the edge of these type of scales have several spines.

Certain modified scales are also found besides the above types. The scales may be elongated to form spines as in porcupine or these may be like sting as in sting ray.

The scales are a small portion of the total length of the fish. If in a year the fish grows half in former length; the scales will show an annulus about half as broad as the total of the earlier annuli. In some cases, the scales are never developed or they may remain hidden in the skin, as in eels. Since the scales grow with the fish, and the fish grows unequally in different seasons, the markings on the scale have been used to determine the age of the fish. In spring and summer, fish grows quickly and thus, a large number of rings or annuli are laid down in the scales. But as the growth slows down in winter, the annuli become less in number and closer together. These rings only represent the physiological condition of the fish.

1.1.4. Coloration

Fishes exhibit an almost limitless range of colors. The coloration in fishes may be grouped into categories - background color and true coloration. The background color may be due to underlying tissues and body fluids, but the true coloration is mainly due to skin pigments. Generally the fish body backside background color is darker than the ventral side with gradual shadings on the sides. Of course, in gold fish (*Carassius auratus*) an almost uniform bright color is observed all over the body. A number of freshwater species are beautifully colored. *Colisa fasciata* is greenish blue above, dirty white below with blue spots on the operculum and colored bands on abdomen. The fins are orange or red in color. A siluroid fish (*Mystus vitattus*) has golden color with bluish shoulder spot and three to four longitudinal dark bands on each side of the body. *Labeo rohita* is blackish black on the dorsal surface, reddish along the sides becoming silvery below. Maheseer (*Tor tor*) is dark grey on the back with reddish or golden sides, silvery abdomen and reddish or yellow fins. Some coral reef fishes are very brilliantly colored. The ornamental fishes, particularly of marine environment, are also having beautiful coloration. Fishes inhabiting in weed infested water bodies or in coral reefs are generally brownish or yellowish dorsally and they possess transverse strips or patches on the sides.

An adaptation is furnished by the coloration of fishes. The color of most fish matches that of their environment i.e. the habitat and depth at which the fish lives. For example, most fish that live near the surface of the open sea have a blue black or green dorsal color and have silvery color on the ventro-lateral sides, which matches the surface water color. The bottom dwellers are darker in color. Hill stream fishes are mostly striped or spotted. Many fishes are brightly colored for territorial advertisement or as recognition marks for other members of their own species. The coloration helps the fishes to hide from enemies. Many fishes can change their color, by expansion and contraction of the pigment cells known as chromatophores and guanophores. The former lies beneath the epidermis of the skin, above or below the scales. These two cells are present in different proportions in different species and in different body parts.

Chromatophores are branched connective tissues. The different color producing chromatophores are with different granules, which may be carotenoids (red or orange). These are called erythrophores. Similarly the melanophores have melanine granules (black) and xanthophores have flavin (yellow). Other pigment granules are purine (white or silver), pterins, porphyrins and bile pigments. The combination of the above three chromatophores produces other colors. In guanophores, there are a number of small grey plates. These are very opaque, and have a strong reflecting power. So, they are called as iridocytes. According to the way, the light is reflected from them, they appear white or bright silver. They contain a crystalline waste product, guanine, which gives the characteristic color. Thus, in the cod, the darker color of the back is due to the great abundance of chromatophores, which come under the influence of light ray. The coloration of fishes is basically of two kinds -

1. Schemachromes coloration as found in different internal and external body parts such as in eyes, scales, gas bladder etc.
2. Biochromes coloration or true coloration, which is due to the presence of the above mentioned pigments.

1.2. SHELLFISH (DECAPODA)

The shellfish are edible aquatic invertebrates of phylum Arthropoda belonging to class Crustacea, sub-class Malacostraca and order Decapoda. Their body is covered with a protective hard shell or carapace or exoskeleton made up of thick, rigid, chitininous cuticle generally strengthened by impregnation with calcium salts. The outer skeleton is not continuous, but made up of divided sections called somites. The shellfishes are economically important cultivable species. Even though they are invertebrates they are referred as fish as they are cultured like that of fish. Because of the presence of the outer exoskeleton, they are called shellfish. The shellfishes include shrimps, prawns, lobsters, crabs etc. Mollusks are also included under shellfish. However, only the decapods are considered and described now.

The common characters of shellfish (decapods) are -

1. These are crustaceans may be from marine, brackish- or fresh-water.
2. The body is bilaterally symmetrical, triploblastic and divisible into head (6 somites), thorax (8 somites) and abdomen or tail region (mostly 6 but sometimes 7 somites).
3. A presegmented region (acorn) and six anterior segments or somites unite to form the head, which is often fused with some or all thoracic segments to form cephalothorax.
4. The carapace may be present, vestigial or absent.
5. Head bears larval median eye, which frequently disappears in the adult, and a pair of stalked or sessile compound eyes are present.
6. Appendages are typically 19 pairs, 13 cephalothoracic and 6 abdominal. Appendages vary in number but there are two pairs of antennae, one pair of mandibles and two pairs of maxillae on the head and one pair of appendages on each segment of the thorax and abdomen.
7. Mandibles may be with palp or without palp.
8. Abdomen is devoid of caudal styles but terminates in a telson.
9. The female gonopore is on 6th thoracic segment and male gonopore is on the 8th segment.
10. The zoaea is the first hatching stage. Rarely it is a nauplius.

1.2.1. Body Form and Coloration

The more or less spindle shaped elongated body offers least resistance in swimming. The body of a decapod is divided into several distinct morphological and functional regions called tagmata or somite. The more basal decapods have less tagmatized body plans. Although crustaceans exhibit a great variety of forms, the basic crustacean body consists of a number of segments or somites. These somites sometimes are joined to form rigid areas and sometimes are free, linked to each other by flexible areas that allow some movements.

At the front (or anterior end) of the body there is an unsegmented or presegmented region called acron. In most crustaceans at least four somites fuse with the acron to form the head. At the posterior end of the body there is another unsegmented region called telson. The telson may bear two processes, or rami, which together form the furca. These two processes at the tail end of the body vary greatly in form. In many crustaceans they are short but in some others they may be as long as the rest of the body. The crustaceans in general show great variations in the number of somites and the amount of fusion. In the class Malacostraca, there is a typical body plan: the trunk (which follows the head) is divided into two distinct regions, an anterior thorax of eight somites and a posterior abdomen of seven somites, although as a rule only six are evident in adults.

The carapace is a characteristic crustacean feature, arising during development as a fold from the last somite at the back of the head. It may form a broad fold extending towards the rear over back or dorsal surface of the trunk, as in the notostracan tadpole shrimps, but it often encloses the entire trunk including limbs and gills. In the clam and in ostracods, the carapace splits into two valves giving the animal a clam-like appearance. In some decapods, the carapace is fused shell plates. In many decapods, the carapace is projected forward to form a rostrum, which is often sharply pointed and toothed. The carapace is absent in many members of the super order Syncarina. Barnacles attach permanently to hard surfaces and use their highly modified carapace to form mantle. The mantle secretes calcium carbonate shell plates.

The head contains antennules, antennae, mandibles, first and second maxillae, and compound eyes, usually stalked (Fig-1.3). The thorax or pereon have third maxilliped and first to fifth periopods. The abdomen or pleon have first to fifth pleopods or swimmerets, uropods and telson. In crabs and other carcinised decapods, the pleon is folded under the cephalothorax, filling into a groove in the thoracic sterna, thus remaining almost invisible from dorsal side (Fig-1.4 a). In crabs, the abdomen is very short with an uncalcified, soft sternal region and tightly held against the underside of the body. It is segmented, somewhat triangular and thin. The abdomen is narrower in male but wider in female (Fig-1.4 b & c). The mouth parts and abdomen are visible from the underside of a crab but gills cannot be seen. They are soft structures under carapace. The eyes, which protrude from the front of the carapace, are on the ends of short stalks. The mouth parts are a series of pair of short legs specialized to manipulate and chew food.

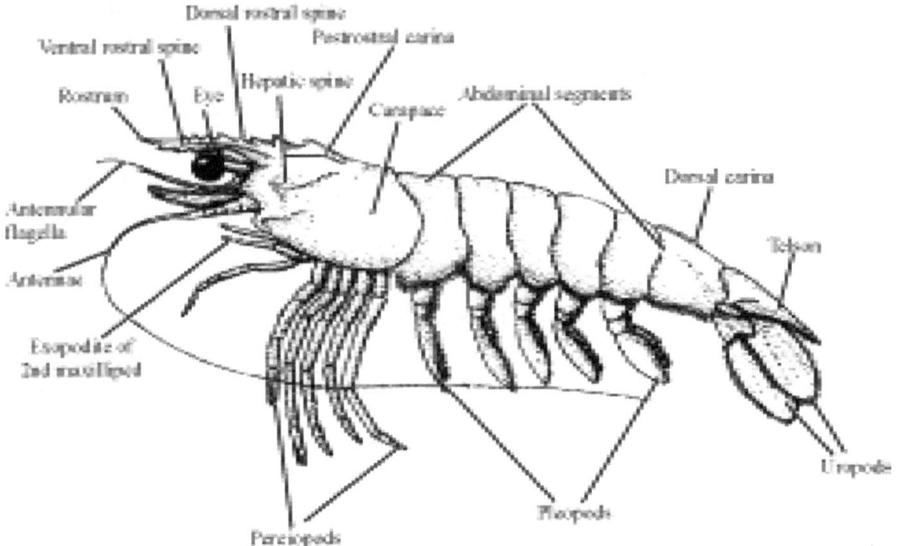

Fig. 1.3 : General morphological features of prawn.

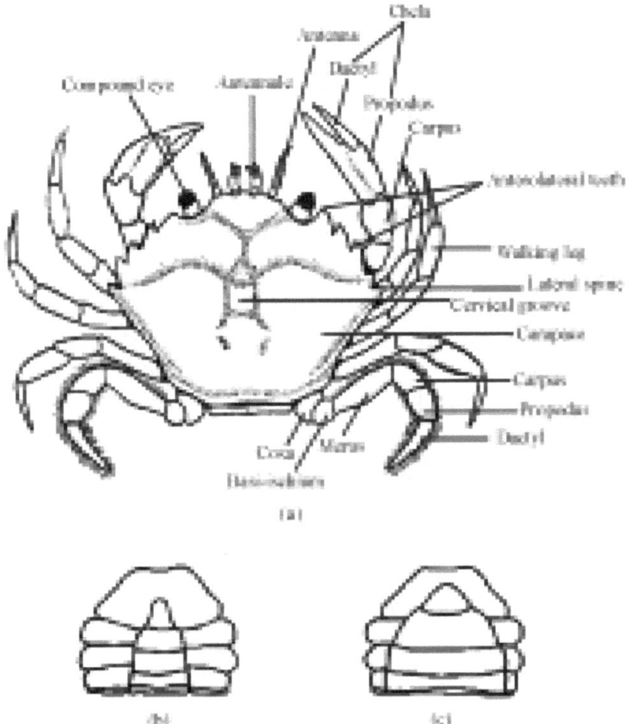

Fig. 1.4 : General morphological (a) features of crab. Abdominal flap of male (b) and female (c) crab.

At young stage, prawns are translucent and white, but the adults are differently tinted according to the species. The usual color is dull, pale or greenish with brown or orange-red patches. When exposed to air, the color changes to yellow or brown. The preserved specimens become deep orange-red.

Decapod's abdomen, which is the edible portion, consists largely of muscles. All crustacean muscle fibers are cross linked. Very little is known about their muscle proteins and the subsequent changes that occur during storage. Near the outer edge of the shrimp abdomen, there is a tough layer (epidermis), which contains much of the pigments, usually melanin and /or astaxanthin (Katayama *et al.*, 1972). The cephalothorax of shrimp may also contain high levels of melanin precursor, which if not removed when melanin production occurs, may stain the meat in the abdomen, and thereby may lower the quality.

The horseshoe crab (*Lumulus sp*) (Fig-1.5) has three main body parts the head region, known as prosoma, the abdominal region or opisthosoma and the spine-like tail or telson. The smooth shell or carapace is horseshoe shaped and is greenish to dark brown in color.

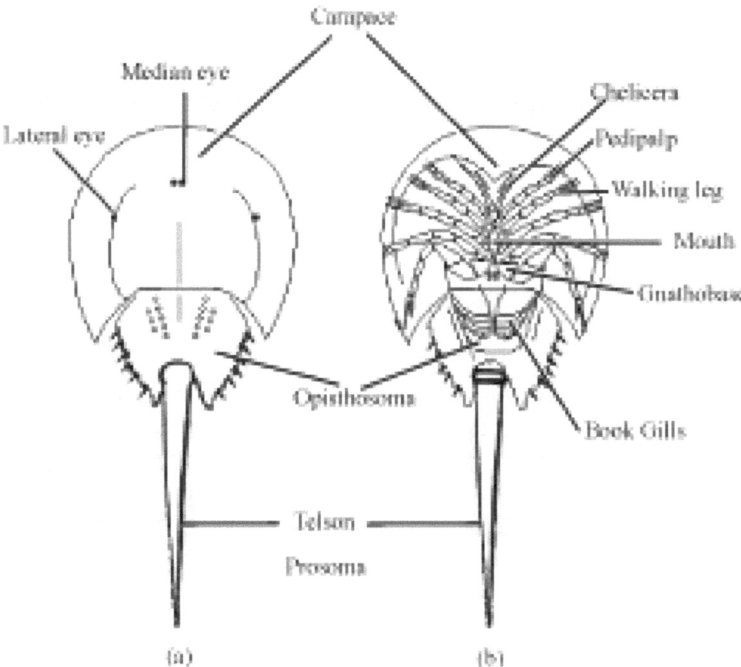

Fig. 1.5 : Ventral (a) and dorsal (b) view of horseshoe crab.

1.2.2. Appendages

The abdomen is enclosed by six segments, placed one after another and connected by a thin membrane. There are five pairs of jointed swimming legs or jointed appendages attached to the first fifth segments, while the legs of sixth segment changed its form into tail fan (uropoda) (Fig-1.3). Above the tail there is a tail that tappers at the edge and known as telson. There is great diversity in crustacean appendages, but it is thought that all the different types have been derived either from the multi-branched (multi-ramous) limbs of the class Cephalocarida or from the double-branched (biramous) limb of the class Ramipedia. A biramous limb typically has a stem or protopodite composed of two segments, namely the proximal coxa for the attachment with the body and the distal basis. The latter bears at its free end two ramis or branches, an inner branch called endopodite and an outer branch called exopodite, each made up of a variable number of joints. The protopodite can vary greatly in its development and may have additional lobes on both its inner and outer margins known as endites and exites, respectively. The first pair of antennae are uniramous in shrimps as their protopodite bears only a single ramous. The walking legs of many Malacostracans have become uniramous by failing to develop the exopodite. With the exception of the antennules all the appendages are homologous, regardless of function due to descent from a common ancestral or fundamental biramous plan.

Starting from the head of a crustacean toward the tail, the appendages are - antennae 1 or antennules; antennae 2 or antennae proper; mandibles, maxillae 1 or maxillulae; maxillae 2 or maxillae proper; and variable numbers of trunk limbs. The trunk limbs all may be similar or they may be differentiated into distinct groups. The appendages are variously modified as jaws, legs, fins, gills and/or accessory reproductive organs. In copepods, the first pair of trunk limbs is used for food collection. These limbs are called maxillipeds. The maxillipeds are followed by four pairs of swimming legs, a fifth pair is sometimes highly modified for reproductive purposes and is sometimes reduced to a mere vestige. In decapods there are three sets of paired maxillipeds.

Behind the maxillipeds there are five pairs of thoracic limbs a variable number of which may bear pincers or chelae or claws. In crabs, there is a single pair of chelae (Fig-1.4 a) and four walking leg. But in some of the prawns there may be upto three pairs of less conspicuous pincers. The decapods abdomen normally contains six pairs of biramous appendages, which are used for swimming in many shrimps and prawns, but in crabs and crayfish the first two pairs in the male are modified to help in sperm transfer during mating. The last pair of abdomen limbs, the uropod, is frequently different from the others. In shrimps and lobsters the uropod together with the telson form the tail fan.

The thoracic maxilliped appendages are modified to function as mouth parts. In more basal decapods, they are similar to the pereiopods. Pereiopods are the walking legs. They are used for gathering food and bear the sexual organs in the fifth pereiopod in the male and the third pereipod in the female. The gills are found in the second maxillae to the fifth pereiopod. Abdominal pleopods or swimmerets are used for swimming, brooding eggs, catching food, sweeping food into the mouth and sometimes bear gills. In some cases, the first one or two pairs of pleopods are referred as gonopods. The tail fan is responsible for steering while swimming. The anus of the decapod is located within the telson.

Under the prosoma there are six pairs of appendages, the first of which (the small pincers or chelicerae) are used to pass food into the mouth, which is located at the middle of the underside of the caphalothorax, between the chelicerae. Although most arthropods have mandibles, the horseshoe crab is jawless. The second pair of appendages, the pedipalps are used as walking legs. In males, they are tipped with claspers, which are used during mating to hold onto the female carapace. The remaining four pairs of appendages are the pusher legs which are also used for locomotion. The first four pairs of legs have claws; the last has a leaf-like structure used for pushing.

The opisthosoma bears a further six pairs of appendages. The first pair contains the genital pores, while the remaining five pairs are modified into flattened plates known as book gills that allow them to breathe underwater, and can also allow them to breathe on land for short period provided the gills remain moist. The telson (tail or caudal spine) is used to steer in the water and also to flip itself over if struck upside down.

The appendages change their form and function during the life cycle of most crustaceans. In most adults, the antennules and antennae are sensory organs, but in the nauplius larva the antennae often are used for both swimming and feeding. Processes at the base of the antennae can help the mandibles push food into the mouth. The mandibles of a nauplius have two branches with a compressing lobe at the base. They may also be used for swimming. In the adult the mandible loses one of the branches, sometimes retaining the other as a palp and the base can develop into a powerful jaw.

One remarkable ability of the crustacea is that they are able to break off or drop their appendages. This is called autotomy. They have special breaking-off points near the body. If caught they can quickly break-off the appendage to escape. A new appendage is more easily grown.

1.2.3. Exoskeleton

The body wall consists of a single layer of glandular columnar epithelial cells called epidermis or hypodermis in which the nucleus is centrally placed. This epidermis is covered externally by a thick and tougher cuticle, and at the innerside of it there is dermis. The non-cellular outer covering of the crustacea is known as integument or cuticle or exoskeleton. It has an outer thin non-chitinous epicuticle layer and an inner thicker chitinous procuticle or endocuticle layer. The epicuticle has an outer thin lipoid layer and an inner thick protein layer. The lipid layer is permeable to gases but impermeable to water. The protein layer is hard and pigmented by its combination with an oxidized phenolic compound. This epicuticle layer prevents the loss and absorption of water. The varying amounts of spines and fringes of setae on the body surface are the outgrowths of epicuticle, which serve as sensory organs. The epicuticle is formed by the secretion of the tegumental glands present in the dermis.

The procuticle is made up of three layers of chitin intermeshed with proteins, and in many species with calcium salts. The outer pigmented chitinous layer have chromatophores and termed as exocuticle. The middle one is calcified chitinous layer and the inner uncalcified plain chitinous layer without any mineral deposits. The procuticle is an elastic layer permeable to gases and other substances. It is formed from epidermis.

The dermis is a thin basement membrane. The dermis is made up of loose connective tissue containing muscle strands and scattered nuclei. It has three types of tegumental glands, each of which individually opens to the outside through a cuticular duct. The secretions of the glands spread out through these ducts on the surface forming the epicuticle. The space between the gut and the body wall is the body- cavity or hemocoel, which is mostly occupied by the muscles and organs with interspaces having blood.

1.2.4. Ecdysis

Crustaceans cannot grow like other animals due to the presence of the hard exoskeleton which prevents any increase in size except immediately after molting. The exoskeleton does not grow. So, they periodically shed the exoskeleton, grow rapidly for a short time, and then form another hard exoskeleton. While this process is taking place they hide in an isolated place. Physiology of molting affects reproduction, behavior, and many metabolic processes. Crustaceans grow in a series of stages or molts. The sequence of events during molting can be divided into four stages.

1. Preecdysis or premolt - This is the period when calcium is reabsorbed from the old exoskeleton into the blood. The epidermis separates from the old exoskeleton, new setae form, and a new exoskeleton is secreted by the epidermis. When crustacean body fills the cuticle, then the animal is in the premolt phase.
2. Ecdysis or the actual shedding of the old exoskeleton - This talks place when the old exoskeleton splits along preformed lines. In the lobster, it splits between the carapace and abdomen, the body is withdrawn through the hole, leaving the old exoskeleton almost intact. In isopods, the exoskeleton is casted in two parts; the front portion may be casted several days after the hind part. Immediately after the ecdysis the crustacea swells due to rapid intake of water.
3. Intermolt or instars-molting—Animals grow in this phase. Soft tissue increases in size until, there is no space within the cuticle. During the molting process, there has been solving the cuticle between the carapace with intercalary celeryte, whereas the anterior part of cephalothorax and appendages are interested to stretch. This phase is of variable duration from a few days in small forms to a year or more in some of the large forms. The molting frequency of white shrimp decreases along with the increase in size. In the larval stage, the white shrimp molts in every 40 hours at 28° C. Juveniles, weighing 1 - 5 grams, molt in every 4 - 6 days. In a 15gm shrimp the molting occurs in every 2 weeks. Molting occurs often in young animals.

4. Metecdysis or post-molt - This is the stage in which the soft cuticle gradually hardens and becomes calcified. The newly formed carapace after molting is very soft and getting more and more hardened to adjust the size of the animal's body. At the end of this stage the cuticle is complete.

Some crustaceans, after passing through a series of molts reach a stage where they do not molt again. This is called a terminal anecdysis. The molting process is under hormonal control.

Chapter - 2

Digestive System

The knowledge of food, feeding habit, anatomical adaptations and the physiology of digestion of the candidate aquaculture species is highly essential to develop cost-effective artificial feed for successful culture practices. The digestion of poikilothermic aquatic fishes and shellfishes is influenced by many factors, and is different from the digestion of terrestrial animals.

2.1. ANATOMY AND PHYSIOLOGY OF FISH DIGESTIVE SYSTEM

2.1.1. Feeding Behavior of Fishes

Based on the feeding habit, the fishes are generally classified as herbivores, omnivores, carnivores and detritivores. Within these categories, the fishes further categorized as euryphagous, having a mixed diet, stenophagous, consuming a limited combination of food types and monophagous, accepting only one type of food. A majority of fishes are euryphagous carnivores, and their feeding mode and food types are associated with the body form and digestive system. Fish generally change their feeding habits depending upon the availability of food. Fishes may be classified as predators, grazers, strainers etc. Plankton is a collective term for variety of marine and freshwater plant or organisms that drift on or near the surface of the water. Most fishes eat the material that is smaller than them. The smallest fishes eat these planktons, which act as their main food and all smaller fishes are the food of the bigger fishes. Largest shark feed on plankton by straining these tiny marine plants and animals from the water. The zoo-planktons, comprise protozoa, small crustaceans, jelly fish, and mollusks, together with the eggs and larvae of many species of marine and fresh water.

Fishes may be divided into gastric (salmonid, catfish, eel, tilapia, groupers, barramundi etc.) and agastric (craps) based on the presence or absence of a true stomach, respectively. One possible explanation for the loss of stomach in some fishes is that they live in a chloride-poor environment. Large amounts of chloride ions for operating a stomach is bioenergetically disadvantageous.

Interactions between the fish body physiology and their environment mean that feeding behavior, whether in captivity or in the wild, is complex and digestion of food is merely the end result of these interactions. Factors that influence this complex process include temperature, light intensity, seasonality, activity cycle etc. All these factors act on internal physiology, e.g. endocrine system, through the animals' senses. Since most aquaculture species are kept in a relatively confined water area, social interactions associated with obtaining food, maintaining space etc. will also impact on feeding behavior.

The stimuli for food are of two types

1. Factors affecting the internal motivation or drive for feeding, including season, time, light intensity, nature of last feeding, temperature and any internal rhythm that may exist.
2. Food stimuli perceived by the senses like smell, taste, sight and the lateral line system, that release and control the momentary feeding act.

Interaction of the above two factors determine when, how and what fish will feed. Increase in appetite leading to higher amount of food consumption and availability of maximum dietary energy for growth, are the two main aspects to optimize production. Diet consumption will not only influence these factors, but also affect both rearing environment and feed management techniques. A positive side effect to maximize food consumption is a minimization of waste, leaching of nutrients and overall pollution.

Another important aspect in feeding is the time of the day. The fishes, which find food by smell and taste (*Ictalurus*) are mostly night feeders. Others like pike (*Esox*) and many predators that feed largely by sight are most active during daytime. Some fishes stop feeding during spawning, e.g. salmons and lampreys. The aestivating fish also postpones the feeding activity.

Feeding behavior is mostly used as an indicator of appetite and feeding response. Vigorous initial feeding activity indicates starvation while reduced initial feeding response is usually associated with health condition. Appetite, i.e. the amount of food eaten voluntarily at one time, appears to be the increase of stomach fullness, although this does not explain the entire appetite phenomenon. Appetite continues to increase for a number of days after the stomach is empty, indicating that additional metabolic or neutral mechanisms are operating, as shown below.

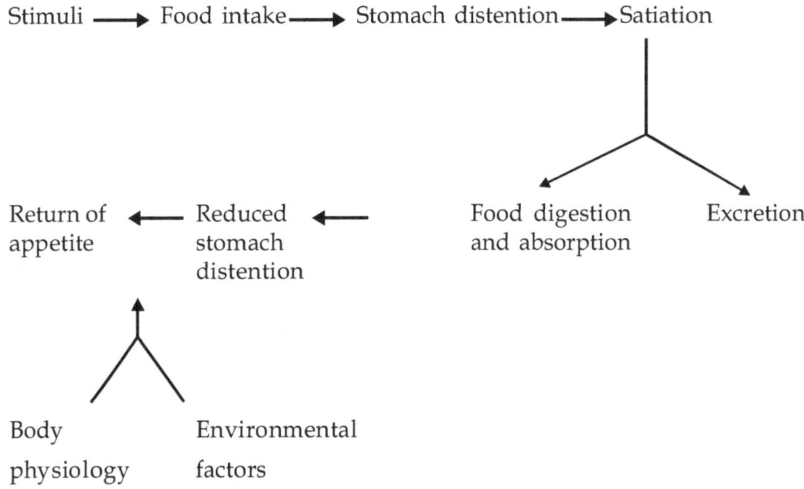

At the physiological level appetite appears to be regulated by the amount of stomach distention and the level of certain nutrients circulating in the blood. Stomach distention acts through the brain to reduce feeding response. As food is digested and passed through the intestine, distension reduces and the brain triggers feeding behavior. Absorbed nutrients from the digested food circulate in blood, and changes in the levels of these nutrients in the body modify feeding behavior.

Metabolism also has an influence on feeding behavior as it appears that food intake by fish is such that it meets their energy requirements. Thus, if a diet has a low energy value, fish will compensate by eating more to a highest level of stomach fullness. The physical limits of a fish stomach means that diet with different energy values, though nutritionally balanced, will produce the same amount of growth.

2.1.2. Anatomy of Fish Digestive System

Irrespective of their main dietary requirements, the digestive systems of different fish groups are very similar even though many modifications are noticed. The digestive system consists of alimentary canal, and its associated glands. The alimentary canal or gut of fish consists of mouth which opens into bucco-pharynx and then continues into esophagus, stomach, intestine and rectum, and ends at the anus. The lips, buccal cavity and pharynx are considered non-tubular part whereas the esophagus, stomach, intestine and rectum are tubular part of the alimentary canal (Fig-2.1). The digestive tube also contains numerous intramural glands, which secrete mucus, enzymes, hormones etc. At the end of the stomach, many bony fishes, have blind sacs called pyloric caeca. These are adaptations for increasing the gut area.

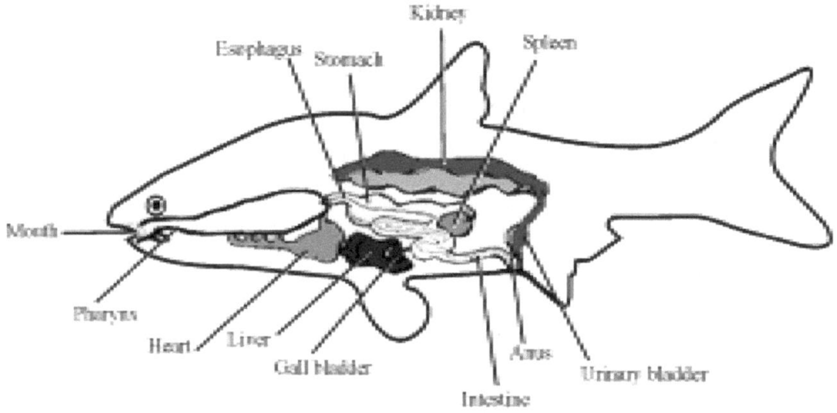

Fig. 2.1 : Digestive system of fish and the associated organs.

The head gut or Kopfdarm is often considered in terms of its two components, the oral (buccal) and gill (branchial or pharyngeal) cavities that are associated with predigestive processing of food. Agastric fishes do not have teeth on jaws. Buccal teeth help only to capture the food. Most agastric fishes have pharyngeal teeth of varying degree depending on the type of food. These teeth can crush the food materials. The filtering apparatus of filter feeders is composed of gill rackers of skeletal gill arches that are modified to form sieve as in silver and bighead carps. Except some carnivorous species, most fishes lack a tongue. Base membrane of the buccal cavity is lined with epithelial line having mucus cells and taste buds.

The foregut and hindgut together are known as Rumpfdarm. The foregut begins at the posterior end of the gills, and includes esophagus, stomach and pylorus. Pylorus is a muscular sphincter which controls the transport of food from the stomach into midgut. Pylorus may be absent in some fishes and in other species it is developed to varying degrees. Where it is absent, the stomach internal lining is roughened to do the grinding work. In the fish where stomach is absent, the esophagus spincter works to pass the food to the intestine. In some fish, like *Cyprinus*, both stomach and pyloric caecae are absent and the foregut consists of esophagus and intestine, anterior to the opening of the bile duct.

The hindgut includes the intestine posterior to the pylorus and a variable number of pyloric caecae or appendages near the pylorus. Pyloric caecae are the extensions of the upper part of the hindgut and found in carnivorous species. They appear not only to be associated with particular type of food and the number, if they are present, but varies from species to species. Caecae are present in salmonids but absent in cyprinids. The midgut is the longest part of the gut and mostly coiled into complicated loops, which are characteristic for each species. In some fish, the beginning of the hindgut is marked by an increase in diameter of the gut to form intestinal bulbs. The posterior end of the hindgut is the anus. Rarely there is a hindgut caecum in fish similar to that found in mammals. Cloaca, a common chamber for anal and urogenital openings, formed from infolded body wall, is not found in teleost fish, except in dipnoi, although it is universal in sharks and rays.

The pancreas may be a discrete organ or it may be diffused in the liver or in the alimentary canal. In elasmobranches (sharks and rays), pancreas is relatively compact, often bi-lobed, and usually well developed as a separate organ. But in teleosts, the pancreas is diffused in the liver to form hepatopancreas. It is diffused in the alimentary canal in a few fishes and sometime present in the mesentric membranes surrounding the intestine and liver.

2.1.3. Anatomical Adaptations of the Digestive System of Fish

In adaptation to diet, the major differences are seen in the structure of the mouth and teeth, gill rakers, pharynx, stomach and in the length of the intestine. The jaws allow the fish to eat a much wider variety of food, including plants and other organisms. The natural position of the mouth is in front but in many fishes the jaws have pulled down and backwards and a snout has grown out above. Thus, in sharks and rays, mouth is ventral and drawn out sideways in a crescent. The cyclostomes differ from all other fishes in having a rounded, funnel like mouth situated at the end of the head. The mouth here acts as a sucker. It attaches itself to other fishes. A mouth which is semicircular and placed on the underside of the head is characteristic of many fishes living in mountain streams. In bony fishes, the primary and both upper and lower jaws are too much modified in the adult fish. In a typical bony fish, the mouth is placed at the end of the head , the upper and lower jaws are equal in length. In some, the mouth lies on the underside of the head as in sharks (Fig-2.2).

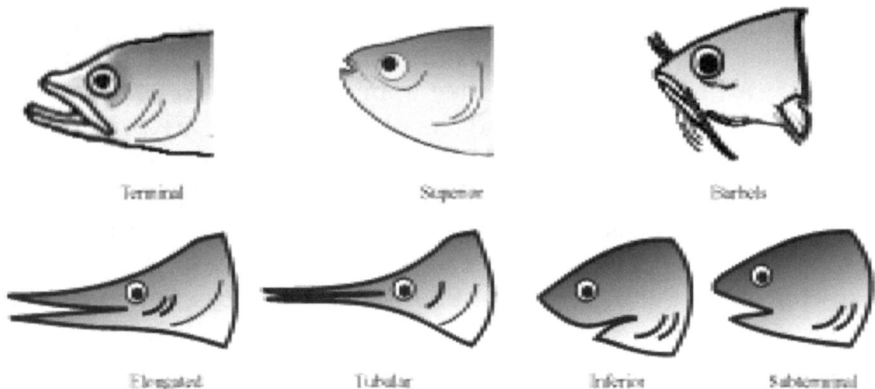

Fig. 2.2 : Types of mouth in different fishes.

In case of sword-fish, mouth is modified into a long snout. Sword-fish attacks on herrings, cuttle-fishes and mackerel with its sharp swings. The pipe fishes have no teeth. They manage to suck the small crustaceans in the water and even a large shrimp. They have elongated beak-like mouth formed by the protraction of the hypo-mandibular bone rather than by lengthening of the lower jaw.

Mostly the jaw equipped mouth has a biting function and fishes that swallow feeders have an inferior mouth and fleshy modifications of lips. The lips of the sturgeons and suckers are mobile and known as plicate with folds or papillose with small tufts of skin. Many suctorial feeders also have well developed barbels more or less bordering the mouth, as in the sturgeons (Fig-2.2). The barbels have many sensory end organs and help to locate food. As a suctorial adaptation, an opercular structure with a separate water inhalent and exhalent device has developed besides the suctorial lips in suckers with inferior mouth.

The pharyngeal teeth of carps and suckers develop from the modifications of the lower elements of the last gill arch. Tooth-like modifications of gill rakers and dermal bones (may be scales) are supplemental adornments found on the inner surface of the pharyngeal arches in many predatory fishes. Some major kind of jaw-teeth is cardiform, villiform, canine, incisor and molariform. Cardiform teeth are more in number, short, fine and pointed. Villiform teeth are more or less elongated cardiform, in which the length to diameter relationship resembles that of intestinal villi. Canines are dogtoothlike. These are elongated and subconical, straight or curved and are adapted for piercing and holding. Incisors are sharply edged cutting teeth. In some fishes they have become variously fused. Molariform teeth are for crushing and grinding and so have flattened often broadly occlusal surfaces. The common carp is an excellent example of non-mandibullar teeth being used as the primary chewing apparatus.

The gill rakers are very stubby and unadorned in omnivores. In many plankton feeders the gill rakers are elongated, numerous and variously lamellated, presumably to augment efficiently in straining. Some gill rackers resemble a feather in having a main axis with lateral processes. When these processes on adjacent rakers overlap, a sieve is formed that can strain.

In most cases, esophagus is short broad and muscular with taste buds and mucus cells. The esophagus in bony fishes is short and expandable, so that large objects can be swallowed. Its walls are layered with muscle. The freshwater fishes have longer esophageal muscles than marine fishes, probably because of the osmoregulatory advantage to be gained by squeezing out the greatest possible amount of water from their food. The esophagus of eel (*Anguila*) is an exception to the general pattern. It is long, narrow and serves to dilute ingested sea water before it reaches the stomach.

The predatory and carnivorous fishes have stomach and the cyprinidae lack the stomach but possess intestinal bulb between esophagus and intestine. According to Dutta and Hossain (1993), 85% teleosts possess stomach and 15% had no stomach. The stomach may be straight with elongated lumen as in *Esox* or U-shaped with elongated lumen as in *Salmo, Coregonus, Clupea* or Y-shaped where the stem of the Y forms a caudally-directed caecum as in *Alosa, Anguila,* the true cods and ocean perch. Fish that eat mud or other small particles have a need for only a small stomach. The Y-shaped stomach is particularly suitable for holding large prey and can readily stretch posteriorly with fewer disturbances to the attachments of mesenteries or other organs. In mullet, *Mugil* and others, gizzard is found which is formed by the thickening of stomach wall. Internally, the stomach is divided into proximal cardiac and distal pyloric region, separated by a constriction in *Protopterus, Esox, Anguila, Raja* and *Mugil*. In some other fishes it is divided into cardiac, fundus and pyloric regions.

The length of the intestine (comprising mid-and hind-gut) varies greatly in fishes. Plant-eating fishes have long, coiled intestine to increase the contact and absorption time. Carnivores have shorter intestine perhaps because meaty food can be digested more readily than vegetable types (Fig-2.3). The absorption capacity of the intestinal area is increased by making its walls into lengthwise folds (typhosole), transverse folds (rugae) and finger-like projections (villi). In elasmobranchs, spiral or scroll valve is found inside the intestine (Fig-2.4), which not only increases the absorptive area but also prevents rapid movement of food materials from intestine to rectum. Spiral valves are also present in dipnoi and some teleosts.

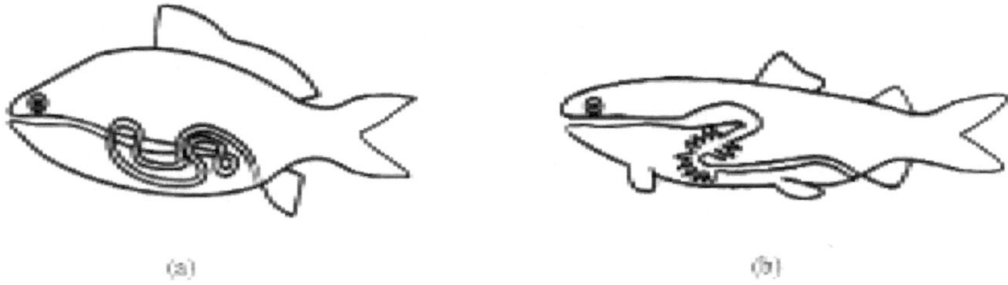

Fig. 2.3 : Variation in the lenth of intestine of omnivore (a) and carnivore (b) fish

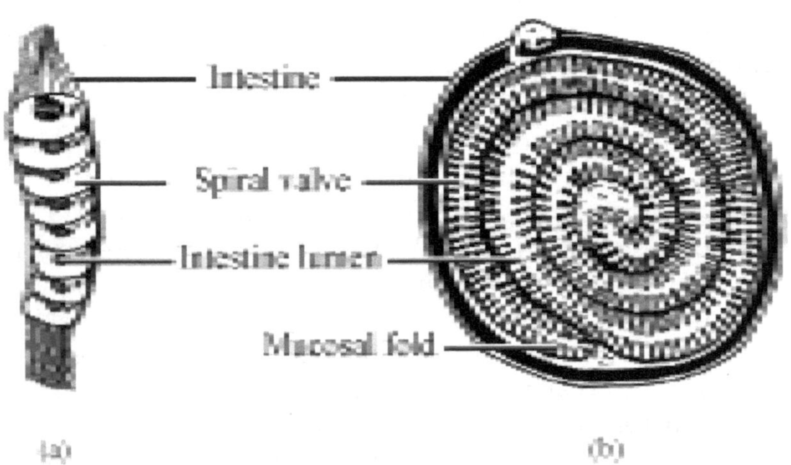

Fig. 2.4 : Diagrammatic representation of intestinal spiral vales, longitudinal (a) and trasverse (b) section.

A useful index which gives an idea of the nature of food consumed is relative length of the gut (RLG). According to the available reports, the length of the gut is almost equal to or in some cases even less than the body length in carnivores but in herbivores it is longer.

Feeding intensity varies from species to species and within the same species from season to season. Due to gonadal maturation, feeding decreases in mature fish, which is termed as "spawning fast".

This happens as gonads occupy major space in abdominal cavity. Food intake in fingerlings decreases in monsoon season due to non-availability of preferred food items. Index of selectivity is a measure to compare a particular item of food found in the gut and that in the environment. It may happen that in spite of the availability of a particular food organism in the medium, the fish does not feed on it but consumes certain other food material. This indicates that fish is selective in feeding. This selectivity can be studies by calculating an index called "forage ratio" or "available factor". This ratio is estimated by taking the ratio between the availability and utilization of a particular food. Preference of a particular food item or the relative importance of various food items encountered in the gut may be expressed in the form of index of preponderance or index of relative importance (Kurian, 1977). In both methods a combination of quantitative and qualitative methods is taken into account for accurate grading of different food types.

2.1.4. Histology of Digestive System

The gut is composed of four layers such as serosa, muscularis externa, submucosa and mucosa. Both serosa and submucosa consist of loose connective tissue. The muscularis externa have an outer longitudinally arranged muscle fibers and the inner circular muscle fibers. The submucosa contains blood vessels and capillaries. The mucosa layer is divided into lamina propria and epithelial layer. The former layer is vascular and is made up of areolar connective tissue. The later one which lines the lumen of the gut, consists of columnar epithelium with many mucosal folds.

The mucous membrane lining the bucco-pharynx is made up of mucosa, submucosa and a thin muscularis. The mucosa contains stratified epithelial cells arranged in several layers, the deepest cells forming the stratum Malpighii. Large sized club-shaped cells are present in the mucosa. At the hinder part of the buccopharynx, many flash shaped mucous secreting cells and a few taste buds are present. The submucosa is vascular consisting of loose connective tissue. The muscularis layer have scattered longitudinal muscle fibers. At the point of gullet, marked folds of mucous membranes can be noticed. In this area club cells are rarely found but large number of mucous cells and taste buds are present. The mucosal folds of esophagus are deep and lined with columnar epithelium containing mucous secreting cells. The submucosa contains bundles of muscle fibers. The circular muscle fiber makes a distinct layer covered externally by a thin serosa.

The muscular wall of the stomach is thick and is lined by columnar epithelium with many mucosal folds. The mucosa contains many gastric glands, which opens to the lumen of stomach. A thin muscularis mucosa is present below these glands. The sub-mucosa is highly vascular. The muscularis has two layers - an outer thin layer of longitudinal muscle fibers and an inner circular muscle fiber layer. The thin serosa is the external covering, which consists of flattened epithelial cells. The pyloric caeca has narrower outer muscular wall than the intestine. Sub-mucosa is almost absent both in pyloric caeca and intestine. When the *stratum compactum* and *stratum granulosa* are larger, then the muscularis is smaller than the stomach. Mucous or goblet cells and endocrine cells are present in a number of enterocytes of the columnar epithelium of both pyloric caeca and intestine. The enterocyte cellular membrane, which borders the lumen is highly folded into a number of microvilli.

The intestine has a thin wall and its mucous membrane forms folds or villi. It contains intestinal glands. In the wider proximal area of the intestine, the villi are more in number. The goblet cells are less and muscularis is thinner in distal part. The mucosa is composed of columnar epithelium having both absorptive and mucous secreting cells. The sub-mucosa is vascular and extends into the villi as in lamina propria. The muscularis is formed of inner circular and outer longitudinal muscle fibers. Serosa has blood capillaries. The intestinal bulb is similar histologically to pyloric caecum. Even though it seems to be a true stomach, its wall is thinner and does not produce acid. The rectum has thicker layers of circular and longitudinal muscle fibers. Mucosal folds are short and flat. The mucous secreting cells are found in large numbers. The presence of neurotransmitter acetylcholine in intestinal bulb and intestine of fish have been reported (Tembhre and Kumar, 1995).

The protochordates have simple, straight gut through which food passes by ciliary action. In lampreys an in folding (typhlosole) of the gut wall presumably increases the absorptive area. Spiral in-folding of the hindgut occurs in sharks, rays and *Latimeria* in the shape of spiral valves (spiral intestine). Teleosts have a gut which is typical of the higher vertebrates in many aspects, although the midgut villi (absorptive papillae) is absent in fish.

2.1.5. Digestive Glands

The extramural glands, which support digestion, beside that of the gut tract, are liver, pancreas and gall bladder (Fig-2.1). Liver secretes bile by the hepatic cells into the bile capillaries and then it is collected in the hepatic ducts, which join to form common bile duct. A cystic duct connects it to the gall bladder.

The pancreas has both exocrine and endocrine part. The pancreas secretes enzymes and bicarbonate into the intestine for digestion. The exocrine tissue produces pancreatic juice, which is carried by the pancreatic duct into the duodenum. It is neutral to alkaline in nature because of the presence of bicarbonate. This juice contains the enzymes for the digestion of proteins, carbohydrates, fats and nucleic acids. The carbohydrate and lipid splitting enzymes are pancreatic amylase and lipase, respectively. Similarly the protein splitting enzymes of pancreas are trypsin and chymotrypsin, both of which are secreted from the pancreas in an inactive zymogen condition. Trypsinogen is first converted to active trypsin by the action of another enzyme, enterokinase, present in the intestinal fluid. Active trypsin converts inactive chymotrypsinogen to active chymotrypsin. Different proteases are secreted either from the intestinal mucosa, pancreas or pyloric caeca. The α- and β-cells of the endocrine pancreas secret the hormones glucagon and insulin, respectively for carbohydrate metabolism.

The gall bladder is vestigial in deep sea fishes but it is prominent in other fishes. It acts as a store place for the bile. Bile contains the fat emulsifying bile salts along with bile pigments, biliverdin and bilirubin, which arises from the breakdown of red blood cells and hemoglobin. This is done in the liver. Biliverdin (green) and bilirubin (red) are the pigments responsible for the color of the bile. As the bile reaches the intestine, changes in the bile pigment take place due to bacterial action. Bilirubin is reduced to mesobilirubinogen, which on further reduction gives stercobilirubinogen. This on oxidation gives stercobilin (brown). This is responsible for the color of faeces. Bile from gall bladder serves to maintain the alkalinity of the gut. Majority of bile salts are reabsorbed from the intestine and returned to liver.

There is no digestive fluid secretion into the esophagus. Gastric glands occur in most of the predatory fishes. These glands secrete gastric juice, which contains HCl and pepsin in inactive pepsinogen form. The pH of some carnivorous fishes is about 2.4 to 3.6. Evidence for stomach enzymes other than peptidases is not clear. Fish gizzard does not have digestive glands. Enzymes lactase and saccharase (invertase) has been found in pyloric caeca in trout and carp, who mainly feed on vegetable matter. Pyloric caeca and intestinal mucosa are sources of the enzyme lipase for fat digestion. Alkaline protease has been found in the intestine and acid protease in the stomach. Small intestine secretes a group of aminopeptidases and dipeptidases (erepsin). The intestinal fluid also contains three carbohydrate digesting enzymes such as maltase, lactase and sucrase. The intestinal proteases are secreted in an inactive form, before chemical changes in the lumen of the intestine making them active for digestion. This change is brought about by enterokinase. This adaptation prevents self-digestion (autolysis) of the intestinal mucosa. In tilapia, amylase activity was found throughout the gastro-intestinal tract but in carnivores, pancreas is the only source of amylase. Fish have endocommensal bacteria possessing cellulase enzyme, which can break the cellulose of plant materials.

Many hormones are secreted from the endocrine tissues of the gastro-intestinal tract. Gastrin is secreted from lower stomach to stimulate the flow of HCl and pepsin from stomach and for the churning action of stomach muscular walls. Gastrin is secreted in response to stomach distensions due to the arrival of food and input from vagus nerves. Enterogastrone, secreted from the intestine inhibits the action of gastrin when there is high level of fatty acids in stomach. Secretin is released in response to acid condition of the gastro-intestinal tract. Somatostain is found in stomach and pancreas of fish. It is a paracrine substance. It differs from hormone as it diffuses locally to target cell rather than being released to blood. It inhibits other gastro-intestinal and pancreatic endocrine cells. Pancreatic secretions are stimulated by another gastro-intestinal hormone, pancreozymin. Cholestochinin another hormone, chemically identical to pancreozymin induces contraction of gall bladder to liberate bile. It is secreted in response to intestinal fatty acids. Pancreozymin is secreted in response to the presence of amino acids and fatty acids in duodenum. This hormone acts back on the stomach tissues stimulating the flow of HCl and pepsin-rich gastric juices. Vasoactive intestinal peptide (VIP) and pancreatic peptides are two other hormones found in fish gastric tract.

Temperature and pH play major roles in the effective functioning of the digestive enzymes. Although most enzyme production decreases at temperatures above or below acclimation temperatures, most enzyme activity increases in proportion to the temperature over a wide range. In general, enzyme reaction rates continue to increase at higher temperature, even though temperature increases beyond the lethal limits for the species, until the enzymes begin to denature at about 50-60°C. On the other hand, enzymes have limited ranges of pH tolerance over which they function, often as little as 2 pH units.

2.1.6. Physiology of Digestion

Before digestion, the food, whether live or pelleted feed, has to be caught, positioned and then swallowed, which is the function of the mouth. In carnivorous or predatory fish teeth do not bite or crush food but simply hold it and prevent its escape. The pharyngeal teeth of herbivores aid in crushing the food before it enters into the stomach. When fishes feed, they maximize their energy intake whilst using the least energy to do so, as optimal foraging theory (OFT) suggests. The ability of the fish to digest and absorb the nutrients in the diet (absorption efficiency) is of considerable importance in aquaculture, where the aim is to produce the largest yield of marketable flesh in the minimum time and cost.

Digestion is concerned with the :

(a) Mechanical breakdown of larger food materials and

(b) Chemical changes of complex food to simple soluble forms.

The digestive system may also function to remove or change dangerous toxic material or properties, respectively of certain food substances. The process of digestion is more or less similar in fish and higher vertebrates. Thus, digestion is a progressive process of hydrolysis of complex macromolecules.

Food passes very quickly from bucco-pharynx into stomach due to secretion of mucous. Since there is no salivary gland in the mouth, there is no digestion in the buccal cavity. In herbivores, the teeth are variously modified for cutting, tearing or grinding. They mechanically break the plant cell walls as the gastric juice is not able to digest chitin. Actual digestion process begins at stomach and ends when food leaves rectum as faeces. The digestibility of the nutrients ranges from 100% for glucose to as low as 5 - 15% for plant materials containing cellulose (fiber) or 5% for raw starch. Digestibility of most protein and lipid materials ranges from 80 - 90%. HCl produced in the stomach facilitates in:

(a) Disinfecting action by killing bacteria.

(b) Converting disaccharides to monosaccharides.

(c) Activating pepsinogen to active pepsin.

Pepsin is a proteolytic enzyme, and hydrolyses the complex food proteins into simple protein molecules or peptons in presence of HCl. Hydrochloric acid is secreted in gastric fishes to reduce gut pH and to allow pepsinogens to work as pepsin. Other enzymes, secreted by the stomach, work on lipids, carbohydrates and chitin. Chitin is the main component of crustacean and molluskan shells that are consumed by the fish. In agastric species, neither HCl nor pepsin is formed in the gut. A study on the digestion in channel catfish showed that digestion and absorption of protein occur during the passage of food through each part of gut (Smith and Lovel, 1973). However, most of the protein digestion occurs in the stomach, but also continues in the intestine. Acid concentrations in the stomach ranges from pH 2 - 4, then becomes alkaline (pH-7.9) immediately below the pylorus, decreased slightly to a maximum of pH 8.6 in the upper intestine, and finally nears neutrality in the hind gut (Page *et al.*, 1976). This, acidic condition of stomach favors the protein digestion in stomach. In agastric fishes there is no acid phase of food digestion. The stomach produces the hormone gastrin, which activates the gastric glands to secrete gastric juice. The gastric juice is a watery liquid, which contains HCl and pepsin besides other substances.

The secretion of HCl and enzyme (s) in teleosts stomach may be by a single type of cell. The production of acid in fishes is due to the reaction of NaCl and H_2CO_3 to produce $NaHCO_3$ and HCl. Blood is the source of input for both the materials, which are later reabsorbed in the intestine. In addition to acids and enzymes, the stomach wall also secretes mucous to protect the stomach from being digested. As long as the rate of mucous production exceeds the rate at which it is washed and digested away, the gut wall is protected from self digestion. When mucous production decreases during gut stasis or stress conditions or post mortem, the gut wall can be eroded or even perforated by the digestive enzymes of the stomach. Both cholinergic and adrenergic nerves are present in the stomach, which stimulate the secretion of gastric juice. The secretion of gastric juice depends upon temperature. At 10°C the gastric secretion increases three to four fold (Kumar and Singh, 1990). The partially digested acidic semi-fluid food mass is known as chyme, which is further neutralized by the pancreatic secretion of bicarbonate, released to upper mid gut by common bile duct.

Pyloric caeca may have digestive and / or absorptive function. Several other functions have been attributed to pyloric caeca including increasing the surface area of the intestine for food absorption, specific site for carbohydrate and fat absorption and adding to the digestive functions of stomach.

Proteins are also degraded in the alkaline medium of the intestine by the action of the enzyme trypsin. As semi-digested food enters into the intestine it is acted upon by the bile and the pancreatic juice, which contains trypsin, chymotrypsin, amylase and lipase. The enzymes secreted by small intestine itself and pancreatic juice are active at pH ranging from neutral to alkaline. Here bile plays an important role to neutralize the acidic condition produced by the HCl of stomach and make the intestinal medium alkaline for the activity of intestinal and pancreatic enzymes. Trypsin and chymotrypsin act on the remaining undigested protein part of the food. These two proteolytic enzymes continue the break down of large protein and polypeptide molecules into smaller molecules. This process is by hydrolysis with the two enzymes, acting inside the protein molecules (endopeptidases) rather than at the ends. Following their action, the exopeptidases, such as aminopeptidase (acting at the amino end) and carboxypeptidase (acting at the carboxyl end) break the polypeptides into smaller units. Finally the dipeptidase breaks the peptides to liberate free amino acids. These are present in pancreatic juice and small intestine.

The amylase acts on starch and glycogen in a similar manner but even more effectively than mammals and converts them to sugar while lipase digests fats into fatty acids. The pancreas is presumably the primary site of lipase production. However, lipase activity has been found in the pyloric caeca, upper intestine. The pancreas appears to be the primary site of carbohydrase (amylase) production although the intestinal mucosa and pyloric caeca represent additional production sites in various fish species. The lipase activity is found in pyloric caeca and upper intestine even though pancreas is the main site of its production. The work of lipase is facilitated by the action of bile. There are no digestive enzymes in the bile. It breaks up large fat globules into very small ones, i.e. emulsification of fat, giving much more surface for the lipase to act upon. Bile serves to neutralize HCl secreted by stomach, and to prevent possible ulceration of the intestine. It ensures that absorbed toxins are returned to the intestine for excretion. The size and fullness of gall bladder is indicative of feeding status of fish. A large, distended bladder indicates that the fish has not eaten for some time while an empty flaccid gall bladder indicates that the fish has recently taken food. Bile is excreted from the gall bladder to the intestine via the bile duct.

The quantitative presence of digestive enzymes seems to correlate with the diet of fishes. Herbivores and omnivores, which have no stomach lack pepsin as a low pH proteolytic enzyme (Kapoor *et al.*, 1975). However, omnivores have amylase activities in the gut many times more than that found in carnivores (Volya, 1966). Initiation of gastric secretions of pepsin and acid in white sturgeon (*Acipenser transmontanus*) larvae was concomitant with the switch from endogenous (yolk) to exogenous (feeding) nutrition (Buddington and Doroshov, 1986).

Besides the role of digestion, the liver also acts as storage organ for fats, fat-soluble vitamins and carbohydrates. It has further important functions in blood cell destruction and blood chemistry as well as other metabolic functions such as production of urea and compounds concerned with nitrogen excretion.

The type and physical state of consumed food varies with species. Salmonids eat relatively large prey and thus, they reduce the prey size layer by layer. Gastric digestion proceeds in a layer of mucous, acid and enzymes wherever the stomach wall contacts the food. In the mid-gut partially digested food is liquified which solidifies again after further digestion and reabsorption of water during the formation of faeces. Stomach of juvenile Pacific salmon from sea contained a thick slurry of pieces of amphipods in various stages of solubilization. Fish whose food contains high levels of indigestible ballast, e.g. common carp feeding on a mixture of mud and plants, probably show minimal change in the appearance or volume of their food while it passes through the gut. Microphagus fish, such as milkfish (*Chanos chanos*) whose food is suspension of fine particles, probably also keep it in much the same condition in the entire digestive tract. In general, it appears that the degree of liquification of food is not same in all fishes as it happens in mammals.

2.1.7. Absorption and Assimilation

After digestion, absorption i.e. passing of the nutritive materials through living cells or tissues within the body, takes place. Since the components of commercial diets are finely ground up and carefully balanced, high absorption efficiency is expected. Absorption efficiencies of between 90 - 95% are found in farmed fishes and this varies significantly between carnivorous and herbivorous teleosts. In carnivores the energy absorption efficiency is about 80% and in herbivores it is as low as 40 - 50%. This is because of the fact that the herbivores consume plant materials in contrast to the animal type of food material of carnivores.

Nearly all organic and inorganic nutrients are absorbed in small intestine. Modifications of the gut facilitate the process of absorption. The lining of the small and large intestine are highly absorptive. The absorptive capacity of the intestine is increased by making its wall into folds and ridges but length-wise rugae or villi with their lymph ducts (lacteal) inside are absent in fish. These folds are covered with epithelium, within which a network of blood capillaries and lymph vessels are present. During absorption, nutrients pass from the lumen of the intestine through the epithelium to the capillaries or lymph vessels by the process of active transport. The active transport involves movement of materials against the concentration gradient with the help of energy from ATP. Some substances, on the other hand, penetrate passively and diffuse through these folds.

For absorption, the nutrient components of food materials should be of small size that will enable them to cross the membranes of the cell lining the digestive tract, pass into circulatory system, and finally be carried to and enter into the cells or tissues that need them for further oxidation to give energy or re-synthesis for growth or store them. Secondly the nutrient components must be in aqueous solution in soluble form for absorption. Fat (which cannot remain soluble in aqueous medium) absorption is intensified in the pyloric stomach of some fishes and in the pyloric caeca of others. Fats enter into the lymphatic ducts in these regions without being split into their component fatty acids and glycerol molecules upon which intestinal absorption depends. Fat absorption from the intestine is not clear. It is shown that long chain fatty acids are absorbed very largely into the lymph, whereas short chain fatty acids enter the portal blood. For absorption of fat, it seems that lipase and bile salts are required. Increased numbers of leucocytes in general circulation following a meal in the sea bream and increased number of fat droplets in them have been reported (Smirnova, 1966). It was postulated that leucocytes enter the gut lumen, absorb lipid droplets, and then return to the circulating blood. Villi, the lacteals, serve as a primary uptake route for droplets of emulsified lipids (chylomicra). Teleosts have a lymphatic system which includes extensions into the gut wall, but its role in lipid intake is doubtful.

Although most digested proteins are absorbed into the intestine capillaries and then enter into portal veins to be carried into general circulation. Some absorption of protein derivatives has been reported to occur in the stomach of shark. Absorption of amino acids, peptides and simple carbohydrates occurs by diffusion or they are more rapidly transported across the gut epithelium into the blood. Brush border on the surface of the epithelial cells facing the gut lumen are micro-villi, which are finger-like projections of the cell membrane. Their increased surface area is probably involved in absorption. The dipeptides of small intestine undergo further intracellular digestion for the final break down of proteins into amino acids. The amino acids absorbed into the blood diffuse into the body fluid, and so reach all the tissues and cells. At the same time most of the tissue proteins are continuously hydrolyzed to release amino acids, which also enter the circulation and this becomes the "amino acid pool". From this pool, amino acids are taken up as much amino acids as it looses. It is the state of dynamic equilibrium. If the loss is greater, the cell wastes and if the gain is greater the cell grows. Amino acids, if used for building new tissue, could be used as absorbed. If amino acids are to be oxidized for energy, deamination (removal of the amino group) must occur first. This process of oxidation and resynthesis of amino acids are in a steady state to maintain equilibrium. This process is known as specific dynamic action.

The absorption of sugars from the stomach and colon is very less. The proximal part of the small intestine appears to be the major site for absorption of sugar. When concentration of sugar in gut exceeds that of the blood, simple diffusion of sugar from gut occurs. But major amount of sugar is transported from the gut by active transport process. The absorption of sugar is dependent on the presence of Na^+ ions, probably on Na^+: K^+ ratio.

Inorganic ions are absorbed from the different parts of the gut in fishes and their subsequent distribution and localization has been reported. The iron (Fe^{2+}) is absorbed through intestinal columnar cells and then passes into the portal blood as Fe^{2+} binding with protein transferitin. The calcium (Ca^{2+}) is absorbed by the intestinal submucosal blood vessels. After entering into blood vessels in the intestinal region, Ca^{2+} reaches finally the hepatocytes where it is stored in association with vitamin D depending on the Ca^{2+} binding proteins. In teleosts, ambient water also serves as external source of various dissolved minerals besides food.

Defaecation often occurs 2 to 3 hours after meal, presumably due to the undigested materials from the previous meal. After a single meal, faeces are found 24, 48 and 96 hours after the meal. Thus, there is considerable variation in food passage time, which may be related with the digestibility of the food.

Cholinergic and adrenergic intrinsic nerve network controls peristalsis of stomach. The esophagus and stomach are also innervated extrinsically by branches of vagal (cranial X) nerve.

Larger meals are often first, but not always, digested at a faster rate than small meals and the amount of pepsin and acid produced is somewhat proportional to the degree of distension of the stomach. The gastric evacuation rate decreases more or less either exponentially or linearly with time. Variables considered with feeding rate and gastric evacuation time includes temperature, season, activity, body size, gut capacity, satiety and metabolic rate.

2.2. ANATOMY AND PHYSIOLOGY OF THE SHELLFISH DIGESTIVE SYSTEM

The most often cultured economically important decapod shellfishes are prawns, shrimps, lobsters, crayfish and crab. The horse-horse crab, *Limulus polyphenus* is a valuable species for biomedical research. The clotting reaction of this animal's blood is used in the *Limulus Amebocyte Lysate* (LAL) test to detect bacterial endotoxins in pharmaceuticals and to test for several bacterial diseases. LAL is obtain from this animal's blood. Enzymes from horse-shoe crab blood are used by astronauts in the International Space Station to test surfaces for unwanted bacteria and fungi. A protein from horse-shoe blood is also under investigation as a novel antibiotic.

2.2.1. Feeding Behavior

The decapods exhibit a wide range of feeding habits and diets, but the greatest numbers of them are of predaceous feeding with scavenging nature. The relative importance of the two habits varies with the species and also with the available food resources. Large invertebrates are their common prey. The food of shellfishes varies, some eat plants or flesh, while some others feed from the bottom of the ocean or anything they can find. The way in which they feed themselves also varies. The cleaner shrimps feed on the mucous and parasites covering the skin and gills of fish. Shrimps (*Macrobrachium or Palaemon and Penaeus*) need food with protein content ranging from 35% to 45%. They are carnivorous animals that feed on small crustaceans, amphypoda and polychaets. Zooplanktons, especially copepods make up the diet of most pelagic shrimps. Shrimps are naturally nocturnal animals i.e. they are active at night to find food and during day time they hide in the substrate or mud. But in pond aquaculture of shrimps, feeding can be done more frequently to increase growth. The crayfish (*Astacus*) is an omnivore, feeding on plant and animal matter either alive or dead. The food is sized by the great chelae and pushed by them towards the mouth or raked in the same direction by the endopods of the maxillipeds.

Crabs (*Brachyura*) use fine hairs on their appendages to filter feed. Scavenging-detritus feeding is characteristic of many crab species. Detritus feeding grades into filter feeding. The fiddler crabs scoop up mud and detritus with cheliped. The third maxillipeds open, and the collected materials are pushed into the mouth. In others the prey (or food) is caught with the cheliped and then passed to the third maxillipeds, which push it between the other mouth parts. While a portion of the prey is bitten or held by mandibles, the rest is torn away by the maxillae and maxillipeds.

The chelipeds of crabs often reflects feeding habits. Species that scrap algae from rocks or feed upon detritus from the surface of sand and mud, commonly have chelipeds with spoon-shaped fingers. The crabs, which consume mollusks in their diet, have dimorphic chelipeds. The right heavier claw bears blunt proximal teeth in the fingers for crushing the food materials and the left slender claw is used to cut the food. The crabs are carnivorous in nature and require high level of protein in their diet like that of shrimps.

Herbivorous decapods of both freshwater and marine, feed on algae. They utilize vegetation in various stages of decomposition and also eat animal carrions.

In decapods, growth is influenced by two main factors i.e. molting frequency and increase in growth rate as shellfishes increase, their body grows during molting. Environmental conditions and food are the two main factors that affect molting.

2.2.2. Anatomy of the Digestive Tract

The digestive tract has more or less entirely characteristic of the teguments of crustaceans. The alimentary system of decapods, which starts with mouth and ends with anus, consists of three parts (1) the esophagus and the proventricularis forming the foregut, (2) midgut (mesenteron) and (3) hindgut or intestine(Fig-2.5). The foregut or stomadeum, and hindgut or protodeum are differentiated very early during embryonic development in most crustaceans. The nauplii larvae after hatching have a non-functional digestive tract with a mouth and the anus opens after 2 to 3 molts. The arthropods are protostome animals in their embryonic development that the mouth originates from blastopore. Gradually the tract develops its own movements and enzymatic activities starts.

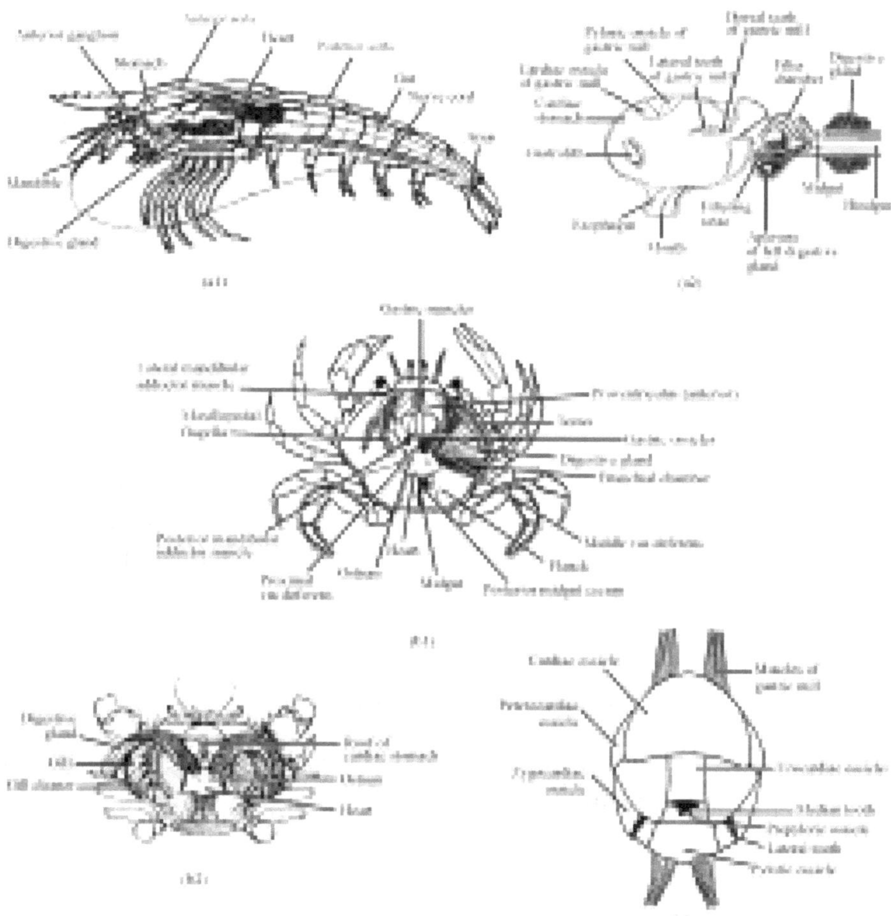

Fig. 2.5 : Anatomical features and digestive system of prawn (a1-a2) and brachyuran crab (b1-b2) and gastric mill (c) of decapod.

The mouth is associated with specialized prehensible appendages maxillula, maxilla, mandibles and maxillipeds. The anterior part of the mouth is covered with a hard labrum. On the lateral side of the mouth there are incisor processes of mandibles and at the back there is bilobed labium. The mouth is large, slit-like aperture lying mid-ventrally below the anterior end of the shrimps. The mouth leads into a short buccal cavity, which is compressed antero-posteriorly, and has regularly folded thick chitinous lining. The molar processes of mandibles lie opposite to each other in the buccal cavity to crush the food.

In decapods, the J-shaped esophagus is short and it joins the mouth with stomach or foregut. It has an anterior roll, which is an extension of the labrum and two lateral rolls. The esophagus have longitudinal ridges or chitinous folds or valves with short bristles, which may restrict the size of the lumen (along with the extrinsic musculature), presumably preventing the regurgitation of intake food. In the Macrourans (lobster, *Homarus americanus*), esophagus is ventral and the cardium tends to be shorter than that of crabs. For the rock lobster *Penulirus argus*, the esophagus is more ventral and penetrates the cardium more posteriorly than in crab or crayfish. The esophagus is associated with the heart region, where it forms a right angle.

The foregut is a double-chambered, chitinous sac that varies greatly among decapods. The anteriormost region, the cardiac chamber or stomach, is a spacious sac in most decapods with a variety of internal structures to facilitate sorting and mastication of the food. The esophagus leads into this stomach. In crabs, the esophagus penetrates the antero-ventral wall of the cardiac stomach. This stomach is separated by a constriction or cardiopyloric valve from a much smaller ventral pyloric stomach, which lies in the posterior half of the thoracic region. The penaeid stomach is the most elongated among the decapods. The stomach pocket has a narrow floor (Fig-2.5).

Both the cardiac and pyloric stomach are made up of varying numbers of chitinous ossicles or plates, which differ in size and morphology. The ossicles are joined to one another by membranous ligaments allowing the movement by extrinsic musculature that controls the action of the foregut. Some of the muscles lie within the stomach wall and others run from the body wall to stomach. The unpaired ossicles are the anterior mesocardiac, found in the roof of the cardiac stomach and the centrally placed urocardiac ossicles. The paired ones are pterocardiac and zygocardiac ossicles. The urocardiac ossicles extend internally to form median tooth. The lateral teeth arise from the zygocardiac ossicles.

A large triangular ossicle is embedded in the middle of the floor of the cardiac stomach. It is called the hastate ossicle. The posterior triangular part of the ossicle is depressed, fringed with setae along its posterior border and touches the cardio-pyloric valve. Each lateral side of the hastate ossicle is supported beneath by a longitudinal cuticular supporting rod. A narrow lateral groove runs along either lateral border of the hastate ossicle. The cuticular ossicle covering the floor of each lateral groove is like an open drain pipe and is called groove ossicle. Each lateral groove is bounded on its inner side by the supporting rod and on the outer side by a long cuticular ridge ossicle. The inner border of each ridge ossicle is fringed all along with a row of delicate bristles, forming comb-like structure and known as combed ossicle.

The cardio-pyloric constriction leads to the pyloric stomach. It is narrow X-shaped and is guarded by four valves. The anterior valve is formed by the posterior part of the hastate ossicle, the posterior valve by semilunar folds of the stomach wall and the lateral valves by the large flap-like posterior ends of the guiding ridges. These valves are fringed with dense setae forming a sieve through which only liquid food or fine food granules pass to the pyloric stomach.

The pyloric stomach is composed of the unpaired pyloric ossicle, which butts against the urocardiac ossicles. Posterior to the pyloric ossicles, there are unpyloric ossicles. A pair of exopyloric ossicle may be present. In filter feeding shrimps, dorsal median projection of uropyloric ossicle projects into the stomach. The median projection is continuous with a complex series of thin chitinous folds that fills the posterior portion of the pyloric stomach.

The proventriculus (foregut) contains a triturating gastric mill in all decapods,(Fig-2.5c), but the degree of development varies. The mill is least developed in shrimps and other forms in which the mandibles and other mouth parts perform relatively efficient chewing action before the food reaches the cardiac stomach. The greatest development of gastric mill is found in lobsters, crayfish and crab, who consume large food. The gastric mill consists of series of calcified plates or ossicles, that are moved against each other by powerful muscle, making an efficient grinding apparatus. A calcareous body called gastrolith occurs in the mill chamber. This disappears at each ecdysis and may serve as a store of calcium carbonate for deposition in the new cuticle.

The lateral walls of the pyloric stomach are thick, muscular and prominently folded inwards so that the cavity of the stomach is divided into ventral and dorsal part. The floor of the ventral part is covered by a median longitudinal $\wedge$-shaped filter ossicle with ridges bearing rows of bristles that form a felt-like covering over the groove. Its side walls are also covered by bristles. The floor of the pyloric chamber leads to the gland filter or ampulla, which is an efficient stainer or filter. It is present at the junction of gastric mill and midgut. The external morphology of the gland filter in decapods is quite variable with the internal arrangement being very similar. The gland filter is mostly elliptical and composed by upper and lower ampullary chambers that are guarded by setae, which form a screen to stain the food.

The midgut extends from the foregut, through the posterior portion of hepatopancreas, into abdominal somites before joining the hindgut.

The midgut is very short but its small surface is compensated by the very large total internal surface of the much branched tubular digestive ceca that opens into it on either side. Anterior and posterior midgut ceca are present in all decapods. The crab has three elongated tubular ceca, two anterior symmetrical ones, the other one is posterior. The lobster has only one short anterior cecum. The posterior midgut ceca extend dorsally from the junction of the midgut and hindgut, and remains freely in the hemocoel.

The hindgut or intestine extends all along the abdomen, from the posterior end of the stomach and terminates generally at the anus. The hindgut of decapods is located in the posterior half of the 6th abdominal segment. It is of ectodermal origin and lined with chitin. Its length is variable throughout the decapoda, being only 1 to 1.5 long. Its anterior swollen muscular part forms an oval shaped gland, called the intestinal bulb or rectum, and bears many internal longitudinal folds. In penaeid shrimps two parallel channels from the hindgut lumen and join with the rectum, where they come together to form a single channel. The proximal third of the hindgut lumen is much larger than the last third and the mucosa forms the internal folds. In penaeids, the hindgut originates at the rectal gland which is dorsally located and consists of rectum and canal. The rectum opens externally at the anus between the telson and the uropods. The anus is a sphinctered longitudinal slit-like opening situated on a raised papilla at the base of the telson.

2.2.3. Histology of the Digestive Tract

In decapods, there is a localized glandular tissue made up of large epithelial cells with clear cytoplasm on either side of the mouth. The esophagus, foregut and hindgut are ectodermally derived with chitinous linings but the midgut is endodermally derived and lined with a non-chitinous columnar epithelium. The lumen of the esophagus is covered by cylindrical basal epithelial cells. The epithelium is covered by a thin hyaline chitinous material. Three types of muscle fibers are present in the connective tissue, enveloping the epithelium. The most internal are dialation, the middle are circular and the outer layer longitudinal. A few rare glandular materials, similar to tegumentary glands are noticed in the esophagus.

The walls of the digestive tract are covered on all its inner surface by exodermally originated chitin-protein coat, which is removed at each molt with the exoskeleton and this participates in the periodic molt also. This coating does not exist in the endodermally derived intestine. The anterior part of the stomach has thin walls. The posterior region of the cardiac and pyloric stomach are reinforced and supported by a number of articulated calcarius pieces, disks and ossicles. They have chitinous coat. The internal surface of the anterior part of the digestive tract has calcified protrusions of skeletal articulated pieces to grind the food. Each piece is moved by individual muscles located outside the wall and controlled by a group of characteristic nerves.

The low to tall columnar cells of the midgut are present on the basement membrane and mostly have a prominent microvillous border, which in many decapods, like *Penaeus* exhibits a glycocalyx. The apical cell surfaces are connected by prominent junctional complexes. The nucleus may be found at the center or base. The rough and smooth endoplasmic reticulum is present at the apical and basal region, respectively. The cytoplasm of the apical portion contains mitochondria, golgi complex and may secretory granules, which are round - to rod-shaped. Midgut cells possess elongated vacuoles called pleomorphic vacuoles. A large acinar tegumental gland complex surrounds the gut at the midgut-hindgut junction at the level where the posterior midgut ceca arises from the gut proper. The gland cells release their contents via ducts that exit at the anterior region of the hindgut. The midgut ceca are composed of tall columnar cells exhibiting a microvillous border lacking a glycocalyx. The nucleus and other cell organelles like golgi complex, mitochondria, rough and smooth endoplasmic reticulum are present at the base of the cells. Cell cytoplasm contains electrondense cellular materials. The unique feature of posterior midgut ceca is the presence of cells within the connective tissue on the hemocoel. These cells contain many myelin-like multilamellar bodies similar to the alveolar cells of vertebrate lungs (Williams, 1977).

Histologically the wall of the hindgut is similar to that of the midgut. But the innermost lining is a layer of cuticular scales or spines and longitudinal muscle fibers are also present in the longitudinal folds. The cuticular modifications always direct their spines in the direction of the anus and presumably aid in movement of the fecal mass towards the anus. The epithelial cells have many mitochondria.

Digestive tract itself can be a source of food or vitamins or perhaps produces digestive enzymes. But, as the food has a very short transit time in the digestive tract, such a direct enzymatic function would be limited.

2.2.4. Digestive Glands and their Secretions

The large orange coloured hepatopancreas is the main digestive gland. It performs the function of both liver and pancreas. Its colour is due to the accumulation of the pigments like carotene, astaxanthine, zeaxanthine. The hepatopancreas in most decapods, is a large bilobed gland occupying a large volume on either side of the stomach inside the cephalothorax. It arises from the anterior end of the midgut and constitutes 2 to 6% of the total body weight. Each lobe of hepatopancreas is made up of tubules associated with connective tissue. The wall of the tubules is made of a single layer of columnar epithelium, resting on a basement membrane. The tubules join to form larger canals, finally forming the two large hepatopancreatic ducts that open into the ventral chamber of the pyloric stomach, just behind the pyloric filter plate. A network of longitudinal and circular muscles has been found in many species. In the crayfish, there is a very small single dorsal cecum pressed closely against the pyloric stomach. In lobsters, there is also a dorsal midgut cecum, but, it is located in the 6th abdominal segment near the junction with the hindgut.

The hepatopancreas arises from the mesenteron of the embryo as a pair of hepatic ceca. The presence of characterized microvilli cells indicates an absorption function. Secretory glands are present that are characterized by vesicles. Four different types of cells are notified at the apex of the hepatopancreatic tubules of higher crustaceans and their specialization increases towards the base of the tubules. The cell types are E-cells or Embryonalzellen, R-cells or Restzellen, F-cells or Fibrezellen and B-cells or Blasenzellen (Dall, 1992; Ceccaldi, 1997). Al-Mohann and Nott (1987) described a new cell type, the M-cell or midgut cell. The E-cells or embryonic cells are small in size and found at the blind end of each tubule and presumably give rise to R- and F-cells. However, the F-cells differentiate into B-cells (Dall and Moriarty, 1983; Ceccaldi, 1997).

The E-cells are more in the mitosis phase. These cells are characterized by a large nucleus with prominent nucleolus, many developing rough and smooth endoplasmic reticulum, very less golgi apparatus and usually lacking a brush border. In crab, the nucleus has a 4 to 5 μm diameter. The R-cells or reabsorptive cells are numerous for storage purpose. These are tall columnar cells with a prominent brush border, centrally located nucleus and a large number of storage vesicles. These cells have microvilli, which allow efficient absorption. Mature cells accumulate lipid droplets, glycogen particles and minerals like calcium, copper, phosphorus, sulphur etc (Hopkin and Nott, 1980). The immature R-cells, located near the apex of the hepatopancreas tubules are clear and of similar size to E-cells with many mitochondria and a well arranged golgi apparatus basally. Their small nuclei, containing less chromatin and endoplasmic reticulum, are limited to some zones only. The fibrillar F-cells have microvilli like that of R-cells for absorption. These are basophilic and larger than R-cells. These cells are found in the central region of the tubules between R- and B-cells. They have basally located nucleus with extensively developed rough endoplasmic reticulum, which gives fibrillar appearance. Mitochondria and golgi apparatus are also numerous. The cytoplasm of these cells contains small vesicles. These cells secrete and synthesize digestive enzymes and store the minerals. The vacuolar B-cells (blister cells) are large secretory cells with an enormous vesicle surrounded by a dense cytoplasm filled with rough endoplasmic reticulum (Gibson and Baker, 1979). A reduced brush border is present. Sometimes entire B-cells are found in tubule lumen. These cells are the primary producer of the hepatopancreatic digestive enzymes. These cells are responsible for intra-cellular digestion and removal of insoluble waste (Dall, 1992). The M-cells are round and present near the basement membrane of the tubules. Their spherical shape and their position differentiate them from the other epithelical cells. These cells may produce cytoplasmic extensions which ramify among neighbouring cells. One distinctive feature of these cells is the presence of spheres, rods and other membrane bound cytoplasmic organells. These cells are involved in the nutrient absorption and they may presumably have also the storage function.

Decapods have different types of digestive secretions, apart from intracellular digestion. In most cases, the secretion is holocrine, that is the entire content are released into the lumen. But merocrine secretions or continual enzyme production by specialized cells and apocrine secretions occurring near the apex have also been described.

The enzymes secreted by hepatopancreas of crustacea are complex proteins that are able to catalyse specific biochemical reactions with very high efficiency and are selective. The trypsin is a major proteolytic enzyme which normally exhibits high activities. Crustacean trypsin can hydrolyse native proteins (Dall and Moriaty, 1983). Chymotrypsin is a serine protease recently identified in the digestive system of prawns (Guillaume, 1997). Carboxy-peptidases A and B are also proteolytic enzymes whose levels of activity vary widely among the different crustacean species (Guillaume, 1997). In lobsters, main proteolylic activity is due to cathepsin, a lysosomal enzyme released by lysis of digestive gland cells. Aminopeptidases and dipeptidases are exopeptidases that have been identified in penaeids (Ceccaldi, 1997). Astacin is a metalo-protease enzyme with a wide spectrum, which has a unique ability to break collagen (Guillaume, 1997). Collagenolytic activity in some crabs is very less. Amylase (α–1, 4 glucanase) is a glucosidase that can degrade α–1, 4 bonds of amylopectin but it cannot hydrolyse α–1, 6 linkage (Guillaume and Ceccaldi, 2001). Amylase activity is less studied in crustaceans. Relatively high amylase activity has been detected in crabs (Brethes *et al.*, 1994), prawn (Omondi and Stark, 1995) and crayfish (Figueredo *et al.*, 2001). Carbohydrases other than amylase have been identified in crustaceans. This suggests that non-starch carbohydrate based substances may also be potential food resources for crustacea. Other carbohydrate digesting enzymes are saccharases, maltases, laminarinases, chitinases and even cellulase in some species. Cellulase (endoglucanase) that hydrolyse β–1, 4 bonds is found in penaeids (Gonzalez-Pena *et al.*, 2002), crabs (Brethes *et al.*, 1994) and crayfish (Xue *et al.*, 1999; Figueiredo *et al.*, 2001). Xylanase (endoxylanase) activity has been reported in some crustaceans. Laminarinases (α -1,3, glucanases) digest the storage polysaccharides or laminarines present in algae, fungi and marine protozoa. Chitinases and chitobiases are mostly found in predatory crustaceans to digest the chitinous exoskeleton of the pray or their own exoskeleton. Oxydases and nitroreductases are reported from the hepatopancreas of lobster (Elmamlouk and Gesser, 1976). Deoxyribonucleases, ribonucleases and phosphatases have been reported in several species. Lipases have been little studied but many esterases were found to exist in each species. Lipid digestion and absorption is greatly aided by the presence of the emulsifying agents. The optimal pH for the activation of digestive enzymes varies. In many cases, it is higher than that of vertebrates. At pH range from 5.5 to 9.0, the enzymes are active. Temperature also affects the enzyme activity.

The digestive enzymes have their originality, for example special endocrine mechanisms, which regulate their synthesis. Peptide hormones are secreted from the entero-endocrine mucosa cells of the stomach (Favrel and Van Wormhoudt, 1986). These hormones are gastrin, secretin, cholecystokinin and gastric inhibitory peptide (GIP). Gastrin and cholecystokinin are found in the eye stalk, near X-organ medulla terminates and medulla externa, and the sinus gland (Ceccaldi, 1989). In penaeid shrimps quantitative and qualitative variations in these peptide hormones have been observed after feeding. Gastrin stimulates the secretion of gastric juice, contraction of smooth muscle of stomach and relaxation of pyloric sphincter. The later promotes the passage of food from stomach (gastric emptying) into intestine. Secretin stimulates hepatopancreas to release bicarbonate to neutralize chyme, release of fat emulsifying agent and inhibition of gastric juice secretion and gastric motility. So, secretin decreases the stomach digestion process.

2.2.5. Digestion and Absorption

Algae, moss, weeds and small aquatic animals like insects, snails, tadpols, fish and debris of the bottom are the natural food of prawns. Even though the crabs are carnivores or predators, their stomach contents have been found to contain plant materials (Tacon and Akiyama, 1997). In natural habitats larval mud crabs usually eat planktons while the main diet of adults and sub-adults are mollusks, crustaceans and dead fish (Hill, 1976). Crustaceans prefer live food and locate pray by sight, touch and chemical means (Baliao, 2000). Feeds provided as complete or supplementary diets for the culture of shellfishes usually contains higher level of animal protein. The dietary requirements of crab are not nearly as stringent as those of most penaeid shrimps, with good growth occurring over a wide range of protein and lipid levels (Catacutan, 2002).

A high digestibility of plant fiber and ash materials of the artificial feed has been observed in crustaceans, particularly crabs (Catacutan, *et al.*, 2003). Current research on the nutritional requirements of crustaceans has been focused mainly on the target species digestive process and ability to hydrolyse, absorb and assimilate nutrients (Guzman *et al.*, 2001).

Stimuli which influence digestive activities may originate from head, stomach or small intestine to regulate the digestion in respective phases. Cephalic phase of stimuli originates from head due to the sight, smell and taste of the food. At the base of the digestive ceca neurosecretory nerve cells, haemocytes and other cells of endocrine functions are present that participate in the stimulation of digestive process.

The food, which consists of animal carrions (dead fish, bivalves etc.) is grasped by chelae and pushed towards the mouth. In mouth the food is held by the third maxillipeds while the mandibles size it and tear off. Inside the buccal cavity, the molar processes of mandibles masticate the food. Food passes from the mouth parts to the J-shaped short esophagus and moves directly into the mill chamber of the foregut. The peristaltic action of esophageal wall and the sucking action of cardiac stomach at the time of feeding facilitate the passage of food through the esophagus and the stomach. The cardiac stomach expands and contracts so that the food is churned and also digested by the action of digestive enzymes. In decapods, food is crushed into fine particles; it passes over the hastate plate by the bristles of the combed plates. Coarse food parts are rejected through mouth. The particle size is selected at filter level which is in the stomach. In the mill chamber the food is further broken down into small particles by the teeth of the gastric mill. The smallest particles pass through the ampullary channels of the gland filter. The semi-liquid and semi-digested food is filtered through the bristles of the combed plates into the lateral grooves below and then the partially digested food is carried into the ventral chamber of the pyloric stomach through the cardio-pyloric valve. Further, it is filtered by the pyloric filtering apparatus to enter into the hepatopancreas through the hepatopancreatic ducts. The larger food particles enter into the channels more directly into midgut for final digestion. The food is continually worked upon by the lateral outgrowths (teeth and ossicles) and return anteriorly to pass through the system again. The liquid phase flows antero-posteriorly along ventro-lateral sides leading to the principal channels of the pyloric stomach. The food particles larger than 1 mm are excluded. The non-digested particles are passed into the hindgut. Water and salts can pass through the epithelium depending on the stages of inter molt (Mykles and Thearn, 1978).

Digestion is extracellular. Only a short region of the gut is lined by endoderm and capable of absorbing digested food. This region is the midgut and its small internal area is compensated by the blind outgrowths into the haemocoel on each side forming the hepatic ceca i.e the hepatopancreas, in which digestion, absorption and storage of lipid, glycogen and a number of minerals occur. It also has a role in distribution of stored reserves during intermolt. The proventriculus and the hindgut are lined by ectoderm, which secretes a cuticle and so do not play any role in nutrient absorption.

The chemical digestion begins at stomach. The transit time of food products lasts some seconds, not even a minute. Food intake induces an increase in digestive enzymes. Around each hepatopancreatic duct, there are fine circular and longitudinal fibers that are responsible for the movement of liquid throughout the organ. Through the ducts, the gland releases its secretions into stomach, and then digestion and absorption occurs. These muscular fibers cause peristaltic and longitudinal movements of each duct, which have small lumens. The amylolytic lipolytic and proteolytic enzymes of hepatopancreas digest the starch, fats and proteins, respectively. Digestive enzymes activities vary with the time of the day and night following internal circadian rhythms, with two peaks per day depending on the phases of the intermolt cycle and also on the phases of vitellogenesis. Larval development stages affect the digestive enzymes, both qualitatively and quantitatively, where amylases are first present at high concentrations then giving way gradually to proteases towards the middle of larval stage. Such a shift from high amylase concentration to high protease concentration is synchronous with the change in food habit from carbohydrate to protein type. Trypsin is the main proteolytic enzyme in crustaceans.

The functions of the midgut are related not only with nutrient absorption but also with osmoregulation and production of the mucous substance (peritrophic membrane) that covers the non-digested fecal material from stomach, then aiding their movement within the tract (Felder, 1979; Hopkin and Nott, 1980). The synthesis and production of this membrane cause a loss in the nutritional energetic costs of crustaceans.

CHAPTER - 3

Circulatory System

Embryologically, the circulatory system is the first organ system to develop in the living animals. The form and function of this system vary considerably among the animals and are shaped by a variety of environmental factors i.e. phenotypic plasticity. Oxygen (O_2) acquisition and carbon dioxide (CO_2) elimination occur simultaneously at the gill site of fish and shrimp. The gases are transported around the body by the circulation of the blood.

3.1. CIRCULATORY SYSTEM OF FISHES

In fishes, heart is simply a muscular enlargement and modification of a part of the main blood vessels, lying between the hepatic capillaries and the gills. Fishes have a closed circulatory system with a heart that pumps blood in a single loop throughout the body. From the heart, blood goes to the gill capillary bed. Then via the systemic capillaries it returns back to the heart. Usually the gills account for approximately 30% of the total resistance to blood flow, the rest being in the visceral and somatic vasculature. Tuna requires more O_2 and the gill area is large with gill resistance up to 57%.

3.1.1. Heart

The heart is small in size on the basis of the amount of blood it contains. It is a modified blood vessel. The fish heart is termed as venous heart as it receives only deoxygenated blood. From heart the blood is sent to gills for oxygenation. After the gaseous exchanges, the blood is distributed to various body parts. The heart is situated immediately behind the gills but in teleosts, it is relatively anteriorly placed than the elasmobranches. In lungfishes, the heart lies behind the pectoral girdle. The heart is enclosed in a pericardium, which is more rigid in elasmobranches.

3.1.1.1. *Structure of heart and blood circulation*

In most fishes, the heart is branchial and S-shaped. It has four parts or chambers, the sinus venosus, ventricle, atrium and bulbus arteriosus (in teleosts) or conus arteriosus (in chondrichthys and dipnoi) leading to ventral aorta (Fig - 3.1.).

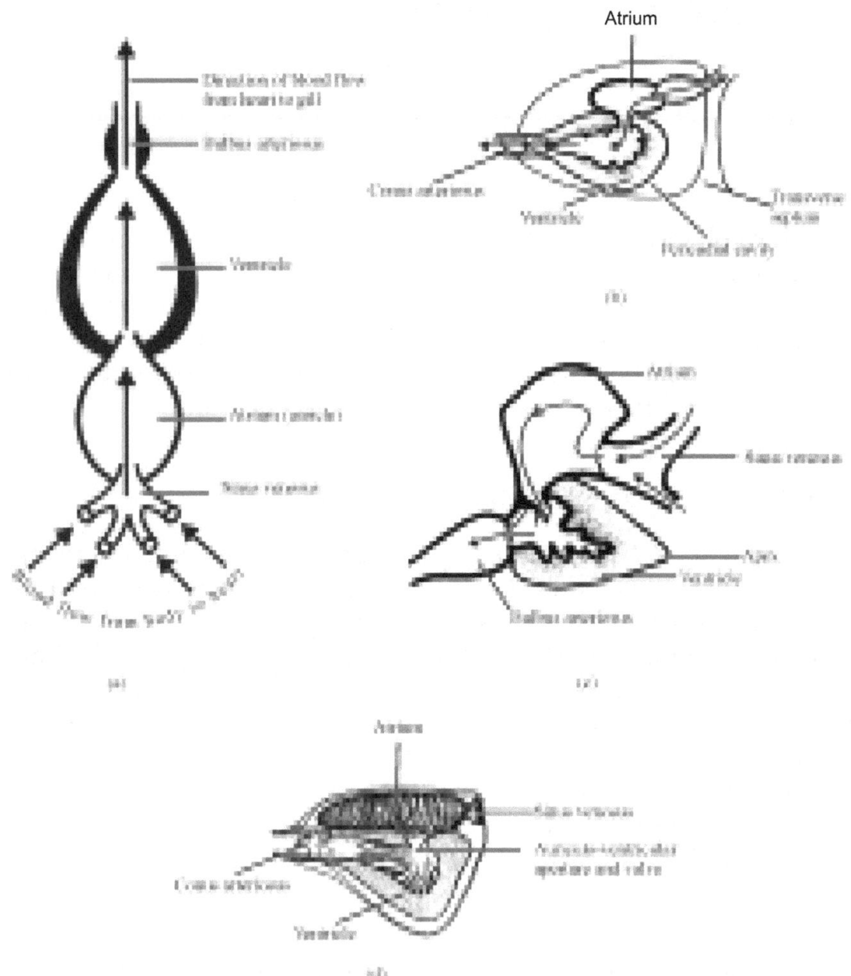

Fig. 3.1 : Diagram of fish heart-General pattern (a), elasmobranch heart (b), teleost heart (c) and sagittal section of elasmobranch heart (d).

All four chambers are in line (or series) and pump only venous blood. In some fishes like Japanese mudfish (*Channa argus*) there are double ventral aorta. Among the four chambers, atrium and ventricle affect significant acceleration of the blood. The ventricle side of the heart is preceded by an enlarged chamber, sinus venosus and the atrial side is followed by a thickened muscular cavity bulbus or conus. Lungfish have a morphologically partially divided heart and the degree of sub-division of heart is correlated with the prevalence of the air-breathing nature of the lungfish. The division is least complete in *Neoceratodus*, who breaths air only when the oxygen content of water is low but other lung-fish *Lepidosiren* is obligatory air - breather. The branchial or systemic heart of fish is the main organ to propel blood. Blood from heart posses first through the ventral aorta into the gill circulation then via the dorsal aorta, it enters into the general circulation. In majority of fishes, the capillary bed of the gills represents a considerable portion of the total vascular resistance to flow and it is in series with other capillary network. Since the blood circulation is linked with gill, the heart is branchial. The circulation is systemic rather than pulmonary. Branchial heart pumps blood from the body and liver into ventral aorta.

Besides the branchial heart, there are other accessory hearts in different fishes for venous blood circulation such as portal, caudal lymph and cardinal heart. The portal heart works independent of branchial heart to circulate the blood to specific area. It pumps blood from gut and anterior cardinal vein to liver. Portal heart contracts while branchial heart is inactive. Branchial and portal hearts are aneural and thus obey Starling's law. The action of skeletal muscle on a cartilaginous plate in the caudal region propels blood forward from large sub-cutaneous sinuses. This cartilaginous plate, the associated skeletal muscles and a pair of lateral sacs with valves, to regulate the direction of blood flow, constitute the caudal heart. The activity of the skeletal muscles of this heart is initiated by the impulses from the central nervous system (CNS). The cardinal heart works through cardinal plexus (or sinus) to supply blood to the head. Hagfish has five hearts besides the usual branchial heart. The five hearts are --two cardinal hearts, two caudal lymph hearts and a portal heart. Portal heart of hagfishes is found behind the liver, which has cardiac-type muscle and resembles the atrium of the main heart. The hemal arch and fin pumps of elasmobranches and teleosts also serve to pump the blood like other accessory hearts. In the sharks, through the hemal arch pump blood from the myotomes is driven into the caudal vein via post ostial valves as the myotomes contract and compress the vascular bed. So, when the fish swims there are cyclical pressure pulses in the caudal vein and the venous blood flows.

Sinus venosus appears to play less or no role in actively moving the blood through the heart but forms a part of an extensive venous reservoir to serve the atrium. It receives unoxygenated venous blood through the paired hepatic veins, a posterior cardinal vein, the inferior jugular vein and the ducts of Cuvier. The openings of these veins are not guarded by valves. The ducts of Cuvier is paired in all fishes but in cyclostomes there is a single such duct in adult. In Indian major carps, the sinus venosus towards large end has a pair of lateral openings. One of the dorsal sinu-auricular valve in rohu (*Catla catla*) extends into the auricle upto the auriculo-venticular apperture and get attached to auricular wall. In *Channa* species, the sinus venous and the sino-auricular valves are absent. The sino-auricular aperture is guarded by five to seven valves.

The sino-auricular valve at the end of sinus venosus opens into the atrium and directs the blood flow with the help of cardiac muscle. While the sinus venosus provides the initial transition from smooth to pulsative flow, the atrium provides the first circulatory acceleration of blood. The atrium has thick muscular walls. It is comparatively larger in size than sinus venous. It receives unoxygenated blood and pumps the same to ventrally placed ventricle through atrio-ventricular and two flap valves. In lungfish the sinus and atrium are dorsal to ventricle. The atrio-ventricular valves are replaced by atrio-ventricular plug. In catfishes, atrium looks like honey-comb and the auriculo-ventricular opening is guarded by four valves.

The ventricle is the largest chamber with very muscular walls. There are two layers of muscles. The cortex is relatively dense cardiac muscle (myocardium), which generally receives O_2 and nutrients from coronary artery. The inner layer of the ventricular myocardium consists of spongy mesh supplied with O_2 and nutrients only by the venous blood it pumps. Many diverticulum are found in the lumen of the ventricle, which are formed by the projections of the inner membrane.

The ventricle tappers at the apex in elasmobranches and takes up conical shape in teleosts (Fig-3.1 b & c). When the ventricle fills with blood it constricts and forces the blood either through the bulbus or conus. The ventricle provides main propulsive force. The cardiac cycle consists of systole, when the ventricle emptied and diastole, when it is refilled. It is accompanied by a progression of electrical waves along the heart resulting from the depolarization and repolarization of the cardiac muscle cells. Thus most fish hearts are myogenic (i.e. no neural input from the brain is necessary for each heart beat) and show a complex electromyogenic wave-form. Contraction of the atrium and ventricle has to be co-ordinated in such a way that there is a delay in the contraction of the two chambers, and there is a rapid synchronous contraction of the ventricle.

Conus arteriosus is a specialized part of truncus arteriosus. The truncus arteriosus of shark and dipnoi is simply an elongation of the conus arteriosus. The border between the two structures is marked, however, by a change from striated cardiac muscle in conus to smooth muscle of truncus and ventral aorta. The conus or bulbus arteriosus does not increase the acceleration of blood. Conus are contracticle but the bulbus is non-contracticle and elastic. It is a blood reservoir that passively enlarges by blood driven forward out of the ventricle. Conus comprise cardiac muscle, but ventral aorta is composed of smooth muscle. Valves at the junctions between the different regions of heart assure unidirectional blood flow, and many valves (up to 72) are also found along the conus. The conus is poorly developed in cyclostomes, which has a single pair of valves. In bulbus there is a single valve to control blood flow. The bulbus wall consists only of elastic tissue and layers of smooth muscles. The bulbus of teleost heart plays an important role in maintaining blood flow and pressure during ventricular diastole. Blood flow due to elastic rebound of the bulbus represents 30% or more of cardiac output. Dampening of flow oscillation in the ventral aorta decreases the work done by the heart in maintaining cardiac output.

The partially divided ventricle of lungfish moves the aterial (from the lungs) and venous (from the body) blood through a bulbus, which largely maintains separation of the two flows by spiral ridges which twist throughout its length. The aterial blood goes to the body, while the venous blood passes through the functional gills, some of it subsequently entering the lungs. The blood from the lungs is directed towards the left side of the atrium whereas blood from the systemic circulation is directed into the right side.

The blood flows from the ventricle into the ventral aorta though the bulbus or conus. The contraction of the ventricle not only drives blood into the ventral aorta but also causes a fresh inflow of venous blood into sinus venosus and auricle. The ventral aorta is either very short or non-existent, and the branchial ateries arise from the anterior end of the conus. Two rows of pocket-shaped or semilunar valves in the truncus arterious prevent blood flowing back into the ventricle from the ventral aorta when the force of the ventricular contraction subsides. From the ventral aorta blood flows to the gills, which have a considerable resistance to the passage of the blood as the blood has to pass there through narrow vessels and the blood pressure falls substantially on the efferent side.

The blood is oxygenated at the gill site. The oxygenated blood is directly distributed to various body parts through the ateries arising from dorsal aorta.

The heart is very contracticle and is supplied by vagus nerves. The contraction (systole) and expansion (diastole) of heart alternates and provides enough pressure (by converting chemical energy into mechanical energy), so that blood contained in it can be pumped to the gills, and then throughout the body. The dissipation of energy as blood flows through the vascular network results in a fall in blood pressure. The highest blood pressures have been recorded from the ventricle during systole (Hoar and Randall, 1970). The resistance to flow between ventricle and ventral aorta is low. Therefore, when the valves are open, pressures in the ventricle and ventral aorta do not differ from each other by more than 1 mm Hg. Intra-ventricular pressure falls when the ventricle relaxes. During diastole ventral aortic pressure falls, valves in the conus (or bulbus) close and pressure decreases as the blood leaves the aorta.

3.1.1.2. *Properties of heart*

The initiation of the heart beat usually occurs in the sino-atrial node. However, the actual site and extent of the pacemaker region varies from fish to fish (Pacemaker is described afterwards). Many fibers in the branchial and portal heart of hagfish have pacemaker potential. The fibers having a pacemaker potential are situated in both the atrium and the ventricle of the branchial heart. In fish many regions of the heart are capable of pacemaker activities. The cardiac resting action potentials of between 40 to 60 mv was recorded in the hagfish portal heart and in atrial and ventricular fibers in the branchial heart of hagfish, elasmobranches and marine teleosts, and above 70 mv in the goldfish and trout atrium and ventricle (Hoar and Randall, 1970).

3.1.1.3. *Cardiac output (Q)*

The teleosts circulatory system is more efficient than that of elasmobranches, blood volume is lower, as is its cardiac output (Q) and narrower veins occur instead of venous sinuses. Antarctic icefish *Chaenocephalus aceratus*, which lacks hemoglobin, Q is much more higher. The Q in a number of elasmobranches is recorded to be 25 ml/kg/min. In teleosts the Q is variable ranging from 5 to 100 ml/kg/min (Hoar and Randall, 1970). But most values fall within 15 to 30 ml/kg/min. The amount of blood entering the ventral aorta is determined by :

1. the volume of blood released from the ventricle at each stroke.
2. the heart rate.

The atrial filling is determined by venous pressure. The ventricle is filled by the contraction of atrium. The atrio-ventricular valves do not open until atrial systole. So, the contraction of the atrium in presence of sino-atrial valves to prevent the reflex of blood into sinus venosus serves to fill the ventricle. Contraction of the ventricle results in large increase in intra-ventricular pressure, the atrio-ventricular valves close and the ventricle contracts isovolumetrically until pressure rises above that in bulbus in teleosts, when the valves at the exist of the ventricle open and blood leaves the ventricle. Blood flow in the ventral aorta falls to zero if the heart rate is low. Usually the minimum flow rate of blood is between 9 and 12 ml/min in a 2-3 kg fish. Intrinsic or resting heart rate varies from around 15 beats/min (bpm) in hagfish to 30-50 bpm in most teleosts and elasmobranches. In resting skipjack tuna (*Katsuwonas pelamis*) this heart rate is about 120 bpm while in swimming skipjack, the rate goes up to 240 bpm (Bone *et al.*, 1995) During swimming cardiac output increases to meet increased oxygen demand of the tissue.

3.1.1.4. Cardiac regulation

The heart rate and stroke volume are controlled by the following.

1. **Pacemaker cells -** The fish heart beats independent of external stimulation due to tightly regulated firing of action potentials from a set of specialized heart muscle cells called pacemaker. This regular firing results from co-ordinated ion movements in the muscle cells. Contraction of cardiac muscle is initiated by chemical impulses. The pacemaker cells create the rhythmical impulses and they directly control heart rate. In fish, the primary pacemaker is located at the base of the sino-atrial valve, where a morphologically distinct ring of tissue comprising myocytes and neural elements are found. Intrinsic beating rate of this pacemaker is higher in cold environment than warm. These pacemakers respond directly to temperature. The cold induced, compensatory increase in heart rate could due to alterations in humoral and neural regulation of cardiac pacemaker activity and /or modifications of the ionic current underlying the regular action potential firing in the pacemaker (Haverimen and Vornanen, 2007).

2. **Autonomic neural control -** The fish heart is enervated by parasympathetic nerve fibers from the vagus nerve. Vagus releases the neurotransmitter, acetylcholin, which decreases the rate and force of cardiac contraction. Vagus activity is stimulated by physical trauma, hypoxia and increase in blood pressure. All these stresses slow the heart rate (brady cardia).

Under resting conditions, vagal tone is low or absent (no release of acetylcholin). The trout heart also receives some sympathetic nerve fibers. These nerves release adrenalin, and could increase the rate and force of cardiac contractility.

3. **Humoral control -** Circulating catecholamines (adrenaline and nor-adrenaline) have great effects on cardiac output, blood pressure and distribution of the circulating blood. Catecholamies may pass into the blood stream either by direct secretion from the chromaffin tissue in which they are stored, or by diffusion from adrenergic nerve endings. These hormones have their effects on the heart and other tissues by combining with specialized regions, the receptor sites, on the membranes of effector cells, which then get activated. Pharmacological studies in mammals have revealed several types of catecholamine receptors like and receptors in the circulatory system. Similar receptor types are also indentified in fish. Generally, catecholamines increase the force and rate of contraction of teleost heart, acting via β-adrenergic receptors. In elasmobranches, adrenaline increases and nor-adrenaline deceases the heart rate, while both these hormones increase the force of contraction. Catecholamines act on α receptors in fish gills to decrease resistance to blood flow. Catecholamines act on α-receptors in the systemic arterioles to cause vasoconstriction. There are few β-receptors in the systemic vascular bed, and there is no dilator response to acetylcholin. Whether adrenergic control of the systemic circulation is via catecholamines circulating in the blood or acting at adrenergic nerve endings is not clear in fish.

4. **Intrinsic control -** The anural branchial and portal hearts obey Starling's Law. This law states that the energy of contraction of heart is a function of the initial length of the muscle fiber. The greater the end diastolic (expansion of ventricle) volume, the greater will be the force of contraction and stroke volume. The venous input pressure to the heart usually determines end diastolic volume. Greater filling of the heart chambers results in greater strengthening of cardiac muscle fibers and more forceful contraction. Starling's Law effect on stroke volume in fish during exercise is less than humoral factors. Starling's Law is very important in balancing the output of the two sides of the heart in mammals.

3.1.2. Blood Vascular System

Since the heart contains deoxygenated blood, it is highly essential to supply oxygenated blood to heart muscle particularly the ventricle, which occurs through coronary arteries (Fig- 3.2 a). Coronaries arise as branches from certain efferent loops. The heart of all fishes has a coronary blood supply. Elasmobranches, teleosts and lungfish (*Neoceratodus*) have coronary vessel originated from the anterior hypobranchial system. But in other lungfish, *Lepidosiren*, the coronaries arise from the second afferent artery (Hoar and Randell, 1970). This transition from an efferent to an afferent origin for the coronary supply is associated with the absence of a capillary network in the second gill arch. This arch mostly contains oxygenated blood returning via the left side of the heart from the pulmonary circulation. Hagfish heart has no coronary supply (Hoar and Randell, 1970). The ventral aorta supplies venous blood to the gills through afferent branchial arteries.

Usually the aortic arches from which the branchial circulation is derived may be consisting of six pairs of vessels associated with the mandibular, hyoid and four branchial arches. The number of these arteries usually vary from four (elasmobranches and teleosts) to five (dipnoi) pairs depending on the number of gills to be supplied. The largest number of seven to fourteen afferent branchial arteries occurs in cyclostomes. In some fishes, the arteries, leading to the gills, arise separately from the main trunk of the ventral aorta. In most sharks the anterior most vessel that arise from truncus, gives rise both to the first afferent branchial artery and to the hyoid artery. Most bony fishes show a common root for gill arteries III and IV. But in carp (*Cyprinus carpio*) gill afferents I and II have a common origin. These arteries curve along the gill arches and farther away they go from the ventral aorta their diameter reduces. In their way they give off arterioles forming capillaries and lacunae in a gill lamellae where the gases and other solutes exchange between blood and outside water takes place. In the capillary bed of the gills reoxygenation of blood and removal of carbon dioxide occur. The blood flow in the gill lamellae is such that reoxygenation and removal of CO_2 occur gradually from aboral to oral edges, in order to make optimal use of the flow of water over the gills. After gaseous exchange the oxygenated blood comes out through a variable number of efferent branchial arteries which again join to form dorsal aorta to supply blood to all body parts. The branchial arteries pass from ventral aorta to lateral dorsal aorta running above the pharynx. In bony fishes (teleosts) one efferent branchial artery is present in each gill arch. But in cartilaginous fishes (elasmobranches), one such artery is for each hemibranch of the gill arch (arteries are paired) expect the anterior hyoid arch which have only one efferent branchial artery. The dipnois show intermediate conditions. Blood pressure drops on the collecting side of branchial arteries in the lamellae of the gills.

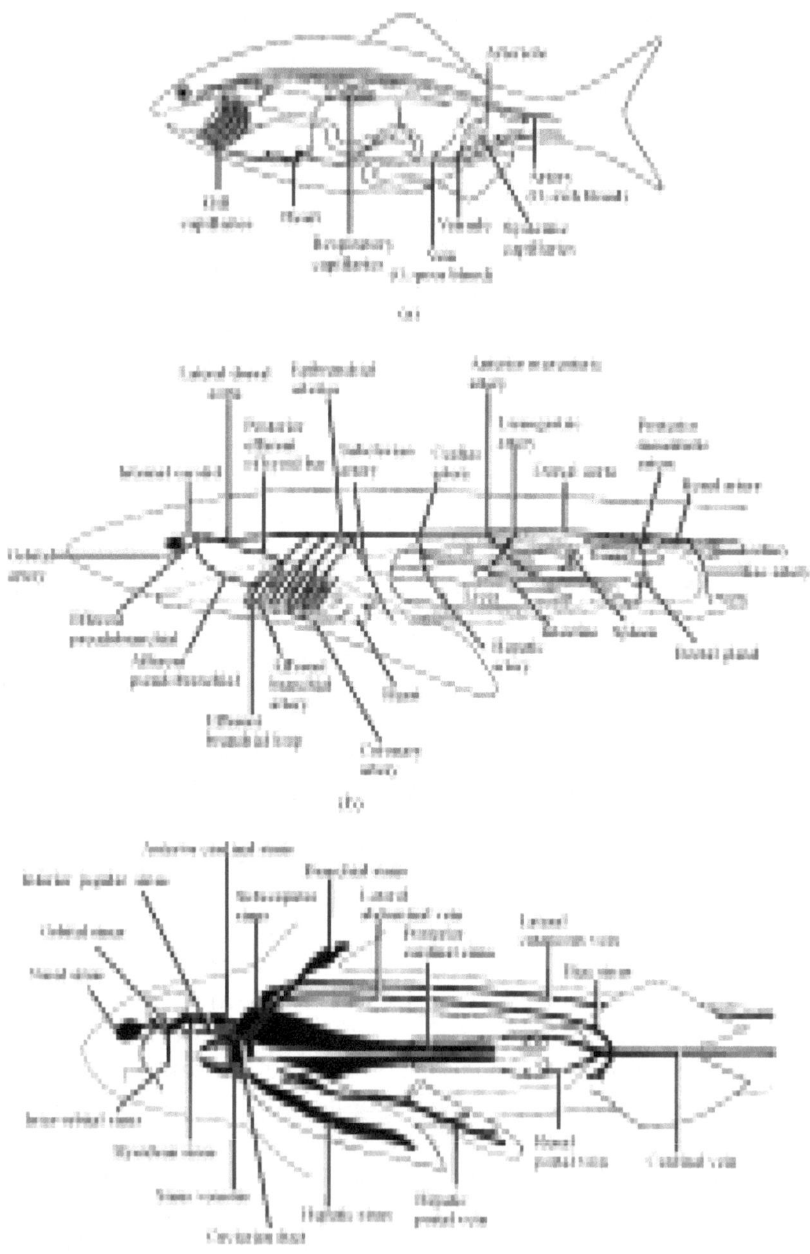

Fig. 3.2 : Blood vascular system in fish, general (a), arterial (b) and venous (c).

In lamprey and sharks the efferent branchial arteries arise dorsally one by one to join the median aorta (Fig- 3.2 b). The first artery mostly provides the blood supply to brain and head region, although some arterial vessels (the hyoid artery) arise from the first afferent gill vessels. In most fishes, a common carotid arises from the first epibranchial artery, which again divides into internal and external carotid arteries. Branches of internal carotid supply oxygenated blood to brain (through cerebral artery), eyes (through optic arteries) and so on. The external carotid gives rise to a number of branches, which carry oxygenated blood to the muscles of jaws (upper and lower) and also to the opercular and auditory regions.

In dogfish, the mandibular arch has no complete aortic arch of its own in the adult but a branch from the first afferent vessel (of the hyoid bar) runs to the spiracle as the afferent pseudobranchial, which breaks up into capillaries in the pseudobranch to reunite as the efferent pseudobranchial. A branch from this forms the orbital artery to supply blood to eyes. The dorsal aorta in dogfish fused posteriorly with the first efferent vessel so that the epibranchial arteries join a single dorsal aorta instead of paired lateral aorta.

From the posterior part of the dorsal aorta a number of arteries arise. Subclavian and iliac arteries supply blood to the pectoral and pelvic fins, respectively. There are small arteries supplying blood to the segmental muscles and four branches from the dorsal aorta go to the viscera. Coeliaco-mesenteric arteries, which further divide into hepatic and gastric, supply blood to liver and stomach, respectively. The anterior mesenteric supply to the intestine and gonad. The lienogastric supply blood to the pyloric stomach and spleen. Several paired renal arteries enter to the kidney and the post-mesenteric goes to rectal gland. The dorsal aorta terminates in the tail region as caudal artery. All these arteries are further divided into small and finer branches.

The arteries have valves to prevent back flow of blood to the aorta. For each segment of the embryo there arises a pair of arteries that divide into dorsal, lateral and ventral branches, uniquely in cyclostomes. Modifications of the embryonic segmental arrangements occur in other fishes as it grows.

There are elaborate sinuses in the venous system of cyclostomes and elasmobranches (Fig-3.2c). This system in hagfish has accessory hearts in the form of contractile bulbs along the way, as the veins of the tails. The main large sinuses are the anterior and posterior cardinal sinuses which release the blood on both sides into the duct of Cuvier lying in the transverse septum, which runs into the sinus venosus. Venous blood retuning to heart flows through sinus-like area of the hepatic and renal systems. Those are the two portal systems. Blood, which is supplied to the gut by branches from the dorsal aorta is collected into the hepatic portal vein and taken to the liver, where again it flows through the capillaries of an organ on its way to the heart via hepatic sinus. The characteristic of a portal system is that it begins and ends in capillaries. The renal portal vein receives blood from the capillaries of the tail muscles via caudal vein and breaking up again into capillaries in the kidneys. The caudal vein in the embryo breaks up in the developing kidney into a network of small capillaries, which then pass the blood into the posterior cardinal veins. The blood thus supplied to the kidneys, together with that supplied by the renal arteries, is drained from the kidneys by the posterior cardinal sinuses.

In shark (*Heterodontus*) there is a gradient in venous blood pressure along the lateral cutaneous vein, from a pressure that is positive and steady at the peripheral end to a pressure that is negative and fluctuating in time with the heart beat at the central end.

Each duct of Cuvier receives subclavian sinus from the pectoral girdle, branchial sinus from the pectoral fin a, lateral abdominal sinus from ventral body wall and a superficial lateral sinus. The inferior jugular sinus from the pharynx joins the duct near the opening of the anterior cardinal sinus.

Some tunas maintain body temperature several degrees higher than that of the surrounding water. This heat is due to the modified circulatory system associated with the red muscle. This is the mechanism of homeothermy in poikilothermic fish like tuna. The mechanism is as follows :

1. As red muscle functions, it generates heat. Muscle-generated heat warms the blood circulating through that muscle, which then travels back to the heart through veins. Thus blood returning to heart from the muscle is warmer than blood travelling from the heart to the muscle.
2. Due to the closeness of arteries and veins, heat passes from warmer veins to cooler arteries within the fish's body, rather than dissipating to the cooler environment. This modified circulatory system retains heat in the red muscle. For this reason freshly caught tuna is warm to touch.
3. A higher body temperature is an adaptive advantage for high-speed swimming.

4. A similar modified circulatory system warms the brain and eye of some species of tunas.

The lymphatic system (Fig-3.3) of cartilaginous fishes consists of lymph vessels. They lack sinuses and contractile lymph hearts. But these are present in bony fishes. In fishes lymphatic fluid (lymph) is collected from all parts of the body by a system of paired and unpaired ducts and sinuses that finally return to the main blood stream. Lymph nodes are absent in fishes.

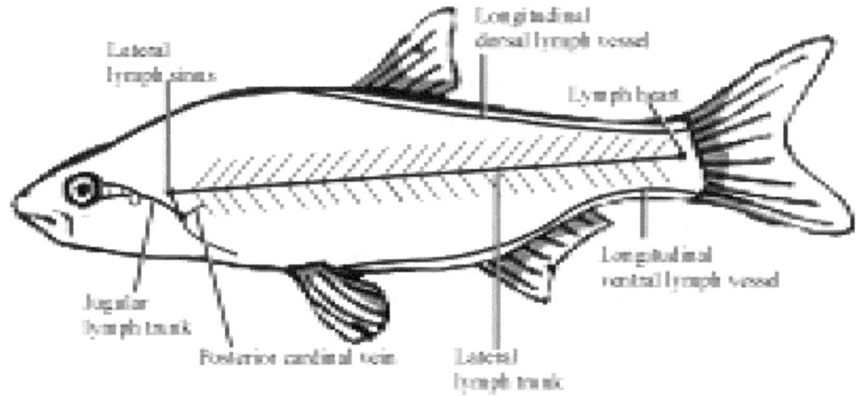

Fig. 3.3 : Diagrammatic presentation of lymph vessels and lymph heart of fish (Based on Hoyer and Bronn, 1938).

The lymphatic system of fishes is originated from the venous system. It is much more complex in teleosts than elasmobraches. The teleosts have well developed lymphatic system like the terrestrial vertebrates. Highly branched subcutaneous lymph vessels are there in teleosts. Lymph from the cranial and branchial regions is collected by branching sinuses and pour into a subscapular sinus in the pectoral region where it is joined with the fluid from the three main lymph ducts of the body such as lateral, dorsal and ventral subcutaneous lymph trunks. Subvertebral lymph trunks, situated in the hemal canal of the tail vertebrae, collect the lymphtic fluids from the tail region and then join into the abdominal lymph ducts which form a network of vessels with the renal and gonadal lymphatic systems. Tissue fluids from body muscles are collected by neural, ventral and hemal trunks and then flow into the lymph trunks, which in turn opens into the cardinal sinus at the origin of the subclavian artery from aorta. The lymph ducts of the visceral organs are divided into a superficial and a deep system. The deep systems collect the fat absorbed from the intestinal mucosa (chyle) into a coeliaco-mesenteric lymph trunk that connects to the remaining lymphatics through subvertebral lymph trunk. The pararenal lymph vessels which end in the pericardinal sinus collect the lymph from the gas bladder, the gall bladder, the abdominal part of the kidney and other organs.

The lymph of Actinopterygii ultimately flows into the main blood stream through anterior cephalic lymph sinus, which opens into the cardinal veins. The orifice connecting blood and lymph systems may remain in the jugular veins (in *Pikes, Esox*) or in the posterior cardinal veins (in *Salmo*).

The excess tissue fluid returns to blood through this lymphatic system or through the capillary wall. The capillary wall is made up of a single layer of permeable endothetial cells. Blood gives up O_2, nutrients and other chemical substances to these cells and the cells release water, CO_2 and nitrogenous waste products into the blood. The spaces between tissues are filled with tissue fluid or interstitial fluid. The hydrostatic pressure of the blood in capillaries forces plasma and other fluid to filter out through the capillary wall. The wall of the lymph vessels is also highly permeable and so, they become a very effective medium for transportation of nutrients and chemical substances from blood to the tissue. Lymph constitutes a part of blood plasma with a few cellular components of blood like red blood cells (RBCs) or erythrocytes. In hagfish, the sinus spaces may be divided into red and white lymphatics, the former having high amount of RBC.

3.1.3. Blood

The volume of blood involved in circulation is low as compared to other vertebrates. It is usually 1.5 to 3% of the total body weight of the fish. Blood consists of blood cells and the fluid plasma. The total number of RBC in fish blood is up to 4 million/mm^2. It contains hemoglobin for O_2 transport from the gills to the tissues. The hemoglobin content of blood varies from 14-19% in dogfish (*Mustclus canis*) to 37-79% in a number of marine and freshwater teleosts (Laglar *et al.*, 2003). The O_2 carrying capacity at 95% saturation, iron content of blood and number of RBC in fishes vary with the life stage. In several Antarctic icefishes (of the family Chaenichthyidae) blood hemoglobin is much reduced or completely lacking and there may be no RBC. The nucleated RBC vary in size in different fish species. Elasmobranches have less number but large RBCs compared to teleosts. Active fishes have more number of RBC than sedentary species. Due to small size of RBC the diffusion path length for gas exchange is shorter. If the O_2

demand is high, blood that is low in RBC has to be pumped through the body at a greater rate. Cameron (1975) has demonstrated for three species of freshwater fish that the teleost heart requires up to 4.4% of the total energy of the fish, the number of RBC can have a significant effect on its energy balance, including growth.

White blood cells (WBCs) or leucocytes are less in number (20,000/ mm^3 to 150,000/mm^3) than RBCs with variety of functions such as providing immunity to the animal to fight against pathogens and blood clotting. Diseased fish makes more WBCs to synthesize antibodies. Among the WBCs there are granulocytes, lymphocytes, thrombocytes and monocytes. Granulocytes that make up between 4 to 40% of all WBC (Laglar *et al.*, 2003). Their average diameter is about 10μ, but ranges from 24 to 33μ in lungfish, *Protopterus*. Granulocytes consist of neutrophils, which are more common, eosinophils (acidophils) and basophils are rare in fish blood. Some elasmobranches have another type of granulocytes i.e. heterophils.

The agranular and nucleated lymphocytes vary in size (4.5 m to 12μm in diameter) among the species and their number is 12 x 10^3/ mm^3. It contains mitochondria and ribosomes in its cytoplasm. The main function of lymphocytes is antibody production. Thrombocytes account for about half of all leucocytes in fish and involved in blood clotting. These may appear as spiked, spindle, oval with a single nucleus. The agranular monocytes are less in number. They mostly appear with response to foreign substances in tissues or blood and they serve as macrophages.

Formation of blood cells (hematopoisis) occurs in many organs of fish. At embryological stage the blood cells differentiate from blood vessels. In adults the blood cells may still be formed in the same way besides that of spleen. Granulocytes come from the submucosa of the digestive tract, the liver, and gonads besides that of kidney. In some fishes, below the submucosa of esophagus there is a flat yellow patch of loose connective tissue filled with WBC, which may be produced there itself. Leydig organ also take part in RBC formation.

Blood plasma is a clear fluid devoid of cells. It contains absorbed products of digestion, waste products of tissues, special secretions such as enzymes and hormones, antibodies and dissolved gases. Sedimentation coefficient of main plasma proteins vary among species. The freezing point depression (Δ) which is a measure of the osmotic pressure, of the fish plasma range from 0.5°C for freshwater bony fishes, 1.0°C in some freshwater sharks and marine bony fishes to a peak of 2.17°C for marine elasmobranches. The main plasma proteins are albumin (controls osmotic pressure), lipoproteins (transport of lipids), globulins (bind heme), ceruloplasma (binds copper), fibrinogen (blood clotting) and iodurophorin (binds iodine). The later protein is unique to fish. The protein content of fish plasma is less than higher vertebrates. The glycoproteins in the blood plasma of Antarctic fishes protect the blood from freezing at a water temperature as low as -1.9°C. The two amino acids present in these antifreeze glycoproteins are alanine and threonine at a ratio of 2:1. The molecular weight of these glycoproteins is between 2600 to 33,000d.The major enzymes of plasma are lipase and carbonic anhydrase, which occur more in marine than freshwater species. The electrolytes present in blood (in mmoles/lit) are sodium (180), potassium (4.9), magnesium (3.8), calcium (5.0), chloride (158) and phosphate (3.1).

3.1.4. Oxygen (O_2) Carrying Capacity of Blood

Blood of active fishes have a much higher O_2 carrying capacity than the sluggish ones. In obligate air-breathers blood is less sensitive to CO_2 content. The hemoglobin increases the O_2 capacity up to 40 times. All the O_2 reaching tissues of several Antarctic icefishes, lacking hemoglobin, depend on the O_2 in solution in the blood, which has the same O_2 carrying capacity as that in seawater (around 0.7 vol %) compared to the higher (0.8 vol%) capacity in many normal fishes with hemoglobin. To overcome the low O_2 capacity of the blood, those fishes have relatively large gills, well-vascularized skin for cutaneous respiration, higher cardiac output with large heart and blood vessels. At larval stage the teleosts also lack hemoglobin and depend on cutaneous respiration.

3.2. CIRCULATORY SYSTEM IN SHELLFISH

Many ostracods, copepods and cirripeds have no heart. The blood is in motion due to a blood pump or rhythmic movements of the body, gut or appendages. In the more primitive crustaceans , such as fairy shrimps or stomatopods, the heart is a long tube, with spiral muscles in its wall, and extends almost the entire length of the trunk .There is a pair of ostia in each somite except the last. The ostia are valvular, slit-like aperture, with inwardly directed flaps or valves. The open circulatory system of decapods have been portrayed as poorly designed systems with poor performance characteristics, lacking in adequate tissue perfusion or fine control mechanisms. Open circulatory systems of at least the higher malacostracan have elaborate capillary beds in many tissues. The open circulatory system, an inflexible carapace and a highly variable volume of the stomach influence the blood pressure. The blood vascular system consists of the heart, the arteries, the blood sinuses or lacunae, blood vessels and blood .There are no veins and true capillaries as in vertebrates.

3.2.1. Heart

The single chambered dorsal auxillary heart of decapods is a coffin-shaped muscular vesicle (Fig-3.4). The *cor frontale* or the accessory heart is controlled and integrated with several other systems. This heart

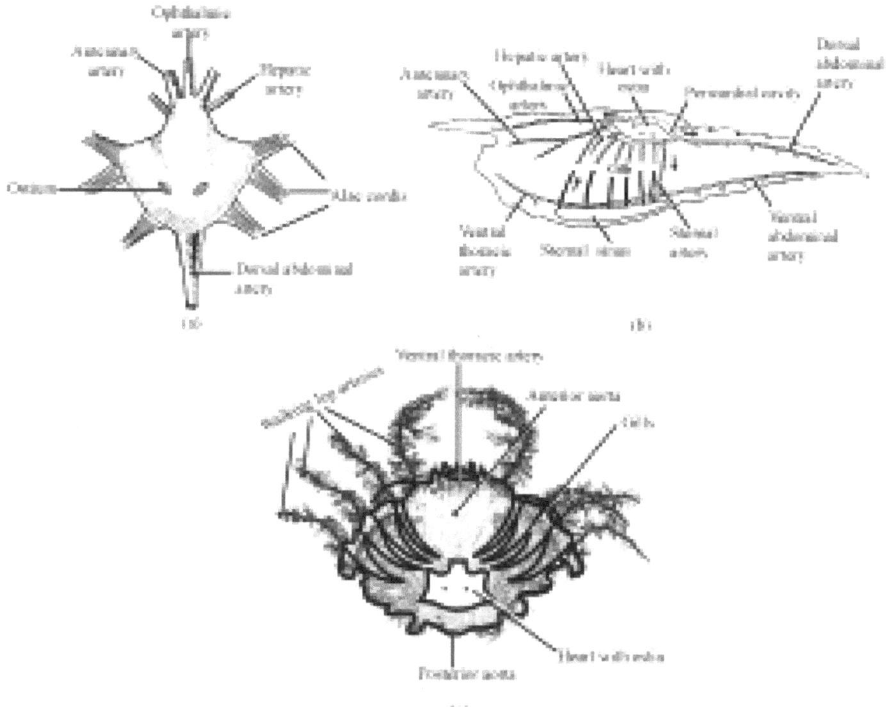

Fig. 3.4 : Shrimp heart (a) and circulatory system (b) and crab heart and circulatory system (c).

is significant due to its adaptation and allows decapods to better maintain blood pressure levels. The position of the heart depends on that of the respiratory organs. It usually lies in the cephalothorax but it is found mainly in the abdomen of isopods. It is located at the end of the dorsal median artery before the artery branches to supply the supra-esophageal ganglion (the decapod brain), and the peripheral oculomotor or visual systems. The heart lies in a space called dorsal sinus or pericardium, which is a part of the hemocoelic chamber situated dorsally at the posterior part of the carapace. The floor of the pericardium is in the form of a thin horizontal septum, lying just above the hepatopancreas and gonad. The septum is attached in front and behind to the dorsal body wall and laterally to the thoracic wall. Most crabs possess a pair of pericardial sacs in the posterior end of the cephalothorax. The sacs are filled with vacuolated connective tissue and some muscle, and are permeated with blood sinuses. One part of the sac is in contact with the pericardial cavity and part of the surface is exposed to the back of the branchial chamber. These pericardial sacs are more extensive and bulge further into the branchial chamber in terrestrial crabs than they do in the marine species. Mostly the gills remain across the bulge of the sac. The function of the pericardial sacs is to regulate the hemostatic pressure of the blood as water is absorbed prior to or during molting.

The heart apex is directed anteriorly and the broad base is posterior. A median longitudinal cardio-pyloric strand of fibrous tissue runs from the apex to the pyloric stomach. Two lateral strands extend from the lateral angles of the heart to the body wall. These strands help the heart to remain in position inside the pericardium. The heart (Fig-3.4a) is provided with three pairs of ostia (mid-dorsal, mid-ventral and posterior), one ostium (anterio-lateral) at each lateral angle of the heart and one dorsal pair (posterior-lateral). In crabs, there are one lateral pair and two dorsal pairs (Fig-3.4c). In more advanced crustaceans, the heart may be shortened and the number of ostia may be reduced to three pairs or less. Blood from the pericardial sinus can enter the heart only through these ostia.

The decapod crustacean heart is considered to be a neurogenic heart (Bullock and Horridge, 1965), since cells of the intrinsic cardiac ganglion initiates the beat. However, nerves from the ventral nerve cord modify both the activity of the cardiac ganglion and the responses of the cardiac muscle. Physiological acceleration and inhibition by regulatory nerves from the ventral nerve cord have been observed. The intrinsic control system comprises an effector (cardiac) muscle, a sensitive brain (the cardiac ganglion) and a sensory input from the stretch receptors on the cardiac muscle. The intrinsic cardiac ganglion has very less neurons, five in shrimps, nine in lobsters and sixteen in crayfish, *Astacus* (Bullock and Horridge, 1965). It is readily accessible in decapods although it lies on the inside of the dorsal surface of the heart, rather than on the outside as in *Limulus*. Decapod heart has a local nervous system comprising of a ganglionated nerve trunk with extrinsic modulator axon from central nervous system. There are also pericardial nerves to the pericardium, the valves and an anterior aortic nerve. The ganglion cells branches so that each motor neuron enters into the muscle cells of the heart wall and ostia.

Each rhythmic cycle of crab cardiac ganglion begins when spontaneous depolarization reaches threshold in one of the small cells (Tazaki and Coo, 1979). This cell actively depolarizes electronically initiating driver potentials in other small cells. In this way depolarization spreads electronically to the large cells and it is reinforced by synaptic potentials from the ongoing brust in the small cells. Conductance to calcium ion increases and may become regenerative as the resting potential of the large cells is depolarized to the normal -55 mv. Lastly voltage dependent increase of potassium conductance determines the peak amplitude of depolarization and initiates repolarization. Each beat of the heart corresponds to an electrical discharge in the striated muscle cells which begins with a large depolarization potential with a steep increase and slow repolarization. Systolic pressure also depends on the degree of contraction of the anterior and posterior cardio-arterial valves, which contract upon depolarization by proctolin (Kuramato and Ebara, 1984). Octopamine hyperpolarizes and relaxes anterior valve cells but contracts the posterior valve cells by depolarization. Regulation of systolic pressure occurs by the neurosecretory pericardial organs, which release serotonin, octopamine and a peptide hormone similar to proctolin.

3.2.2. Blood Vessels

When the heart muscle is relaxed, blood flows into it through the ostia from the pericardial space but when the heart contracts the blood cannot be squeezed back again into this space since the flaps are forced outwardly to close the ostia. Malacostracans have a well-developed system of elastic-walled arteries, including an anterior and usually a posterior aorta. Blood then escapes through the arteries, which lead from the heart to the front and rear of the animal, in a forward direction through the median ophthalmic artery and the two lateral cephalic (antennary) arteries and in a backward direction by a pair of hepatic artery. A median abdominal artery leaves the heart posteriorly and a sternal artery arises either from the underside of the heart or from the base of the abdominal artery. The blood or hemolymph is pumped to the head through an aorta and to the gills and locomotor appendages via lateral and ventral arteries and returns to the heart through a series of sinuses. Several major arteries exist the anterior aorta (or dorsal artery), which may be equipped with an enlargement termed *cor frontale*. The basic anatomy of this modification is remarkably similar among the decapods (Steinacker, 1978).

A slender cephalic or median ophthalmic artery arises from the apex of the heart and runs forward mid-dorsally along the renal sac to join the antennary arteries above the esophagus (Fig- 3.4). On each side of the ophthalmic arteries one antennary artery arises from the apex of heart and runs forward obliquely, passing along the outer border of the mandibular muscle. This artery sends a pericardial branch to the pericardium, gastric branch to the cardiac stomach and a mandibular branch to the mandibular muscle. It then divides into a dorsal and a ventral branch. The dorsal branch further bifurcates to supply the antennules, the antenna and the renal organ. From this dorsal branch an optic artery goes to the eye and it bends inwardly to join the ophthalmic artery of the opposite side to form a circular loop (*circulus cephalicus*). A pair of rostal arteries originates from antennary arteries to supply the rostrum. Hepatic or hepatopancreatic arteries originate from the heart ventero-laterally on either side behind each antennary. These arteries enter into the hepatopancreas within which they further divide. A median posterior artery arises from the postero-ventral surface of the heart. It divides into supra-intestinal or dorsal abdominal artery to supply blood to the intestine up to midgut and the dorsal abdominal muscles, and the sternal artery, which runs downwards through the hepatopancreas on the left side of the intestine to connect with ventral thoracic and ventral abdominal arteries. The ventral thoracic branch runs anteriorly up to mouth for blood supply to the sternal area, the first three pairs of walking legs, the maxillulae. The ventral abdominal branch goes posteriorly up to anus, and supplies to the ventral abdominal area, the last two pairs of legs, the pleopods and the hindgut.

After branching a number of times all the arteries open into the hemocoelic spaces to bathe the various body organs. By its direct contact with the tissues the blood is able to bring to them O_2 and food, and remove the wastes. The blood also passes through the limbs. The arteries enter the ventral sternal sinus, which remains under the endophragmal skeleton in the cephalothorax. From this sinus blood is drawn up by the pumping action of the heart into the gills and after circulating through these, the venous blood ascends by six lateral afferent branchial channels on each side into pericardial space and then goes again to the body. The first channel supplies to the podobranch and the two arthrobranchs. The rest five channels supply to the five pleurobranchs. The aerated blood from gills passes through six efferent branchial channels, which leaves the gills through their gill roots. Although arterial pressures and flows are often lower than those characteristics of poikilothermic vertebrates, the crustacean arterial system is adapted to deliver equivalent flows. The control systems of the crustacean neurogenic heart appear capable of fine graded control over cardiac output. The cardio-arterial valvular mechanisms under neural and neurohormonal control appear to be capable of selective distribution of the cardiac output between the several separate arterial systems. Blood finally drains into a large median sternal sinus prior to passing through the gills and returning back to the heart. Complete circulation has been estimated to take 40 to 60 seconds in large decapods (Barnes, 1982).

3.2.3. Blood

Blood of most mollusks and decapods have copper-containing protein pigment, hemocyanin at a concentration of 50g per liter. Hemoglobin is lacking. Hemocyanin is dark blue when oxygenated but colourless when deoxygenated. Circulating blood in the decapods at cold environment with low O_2 tension is grey-white to pale yellow in color. Hemocyanin carries O_2 in extracellular fluid, which is in contrast to the intracellular O_2 transport in vertebrates by hemoglobin. Blood or hemolymph may be colorless, reddish or bluish in color.

Blood plasma is watery solution of various crystalloids together with colloidal proteins. Hemocytes or WBC or amoebocytes (or ameboid cells) of decapods are circulating cells of the hemolymph that perform a diversity of physiological and pathological functions from wound repair (Fortaine and Lightner, 1973) to clotting of the hemolymph (Stuntman and Dolliver, 1968) to hardening of the cuticle (Vacca and Fingerman, 1983).

Young hemocytes are generally known as hyaline cells and that those cells gave rise to hemocytic developmental series i.e. large and small-granular cells. These granular cells contain a clotting factor, known as coagulogen. Large granular cells mature and accumulate in the connective tissue and can be easily released into the hemolymph while many of the small granular cells are located in the lymphoid organ. Their number further decreases in the hemolymph and increases in lymph node in response to bacterial and viral infections. After bacterial infections, the bacteria accumulates in the lymphoid organ, which has a major phagocytic function. The coagulogen is released to outside of the cells when bacterial endotoxin is encountered. This was observed during white spot syndrome. Many hemocytes are degranulated in the lymphoid organ after bacterial/viral infection, producing a layer of fibrous material in the outer tubular wall. Blood clots quickly on coming in contact with external water. This enables the animal to lose a limb to escape when captured, without excessive bleeding.

CHAPTER - 4

Nervous System

Various activities of the body in different external environmental conditions are co-ordinated by the appropriate message from the neurons due to the stimuli from the environment. These messages pass in the form of impulses through the co-ordinated links between a number of neurons or nerve cells. The constant passage of messages require a very little energy but high level of O_2 .There is interaction between the nervous and endocrine system with the areas of interchange among the two systems. Some neurons show the action of both the systems and thus termed as neurosecretory or neuro-endocrine. Fishes have a well-developed nervous system. Broadly the nervous system can be divided into central nervous system (CNS) and peripheral nervous system (PNS).The brain and spinal cord are the parts of the CNS. The motor and sensory nerves link the receptor and effector organs. All nerve tissues other than brain and spinal cord are grouped under PNS. Peripheral nervous system consists of nerves, ganglia and receptors. The nerves carry sensory information from special receptor organs like eyes, internal ear etc. to the integrating centers of the brain and spinal cord. The PNS also carries information via different nerve cells from the integrating centers of the brain and spinal cord. This coded information is carried to the various organs and body systems, such as the skeletal muscle system for appropriate action in response to the original external and internal stimulus. The PNS can be divided into somatic and visceral. The term visceral is sometimes used in referring to the autonomic or involuntary nervous system (ANS), controlling visceral functions. The ANS helps to co-ordinate the activities of many glands and organs, and it is closely connected to the integrating centers of the brain.

4.1. NERVOUS SYSTEM OF FISH

4.1.1. The Neurons

The functioning of the nervous system is based on the electrical properties of its functional unit, the neurons or nerve cells, which depend on their own intrinsic membrane properties to carry the information in the form of impulses. The changes in these properties are brought about by the electrical and chemical synapses with other neurons. Neurons are quickly stimulated, and transmit impulses very fast. The neurons are joined to one another in the form of a chain. The transmission of impulses from one neuron to the next occurs due to the formation of synapses. There are two synaptic regions, pre- and post-synaptic, where one neuron links with the previous and next neuron, respectively. The impulses are transmitted due to the potential difference between inside and outside of the neuron. At the resting stage the potential is approximately 70m volts.

A typical motor neuron (Fig-4.1) has a nucleated cell body known as soma or perikaryon, containing rough endoplasmic reticulum (ergatoplasm) and free ribosomes. Other organelles of cell body cytoplasm are golgi, mitochondria, lysosomes. Perikaryon cytoplasm and dendrites also contain neurofilaments and neurotubules. Neurofilaments are also found in axoplasm besides that of mitochondria. The elongated tail portion of the neuron is known as axon.

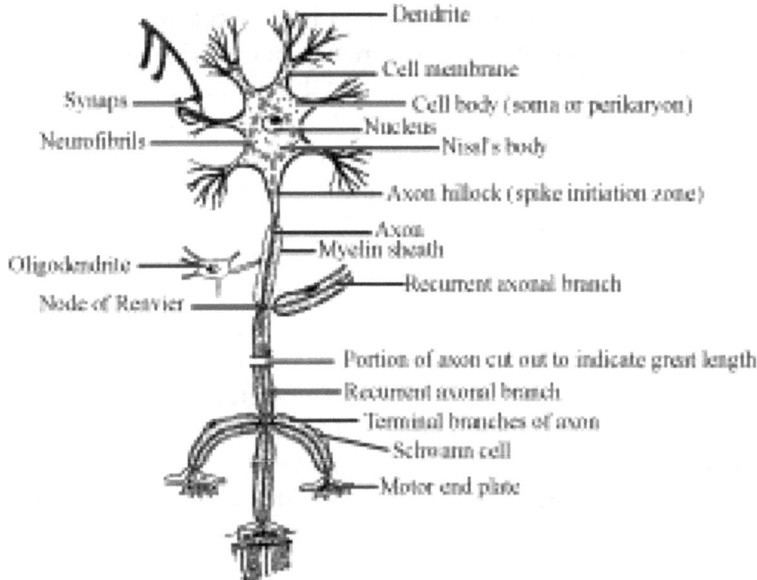

Fig. 4.1 : Diagram of typical motor neuron.

The neurons have two characteristic features such as they never divide and second is the presence of Nissl's bodies. During the degeneration of the cell body of the neuron, disorganization of the Nissl's bodies occur by a process called chromatolysis. Many enzymes and other molecules are present in the cytoplasm of the cell body. Through the process of axon transport these enzymes and molecules are transported.

The cone-shape axon hillock is the site for spike initiation. The information from the external stimuli is received and processed in the cell body and discharged to the axon hillock. The action potential (spike) is generated and is carried downwards by the axon to be released as transmitter substances. The axon is covered by myelin sheath that contains Schwann cells. In between the myelin sheath there are small gaps at some points, which are known as node of Renvier. If by any reason any portion of the never fiber or axon is degenerated, the impulses are released through these nodes that are received by nearly placed other neurons, so that the nervous system functions normally. The intermodal segments are often longer in CNS of fish.

According to the function performed, the neurons are classified into sensory or afferent neuron; motor or efferent neurons and interneurons or connectors or association neurons. The sensory neurons carry impulses from the sensory organs to brain and spinal cord (CNS), the motor neurons carry impulses from CNS to effecter organs such as muscles and different glands, and the interneurons connect the sensory and motor neurons.

Epineurium is the dense connective tissue layer with many blood vessels that surrounds the nerve. This tissue layer penetrates the nerve to form the perineurium that surrounds bundles of nerve fibers. The endoneurium consists of a thin layer of loose connective tissue surrounding individual nerve fiber.

4.1.2. Mauthner Neuron

Actinopterygii (bony fishes) contain a pair of very large multipolar neuron, known as Mauthner neuron in the medulla at the level of cranial nerve VIII, which is absent in elasmobranches (except at the young stage). The lateral dendrites of these neurons join with the cranial nerves V, VII, IX, and X and to the cerebellum and the optic tectum. Their axons pass through the spinal cord to the muscles of the tail and are best developed in active swimmers like salmons and trouts. In eels (Anguilliformes) and molas (*Mola*), these neurons are absent. Input comes from the opposite Mauthner neuron and neurons associated with a special axon cap. The Mauthner axon forms

Mauthner chiasma crossing each other. These neurons are motor co-ordinators for relaying multiple sensory impulses mainly from lateral-line centers to the caudal and abdominal swimming muscles. In lampreys, the oculomotor levels of the brain is co-ordinated with the tail by the neurons of mueller and also cells of mauthner.

4.1.3. Other Supporting Tissues

Neuroglia or simple glia are the supporting cells of CNS, while Schwann cells and satellite cells of ganglia are in PNS. The neuroglia are classified as astrocytes, oligodendrocytes and microglia. In the spinal cord of elasmobranches many branched astrocytes are found. In the cerebellum and spinal cord of bony fishes various neroglia elements are also differentiated. In addition to neurons and glia, there are many blood vessels both in CNS and PNS. The ectodermal epenchymal cells line the cerebrospinal canals in all fishes. These are concentrated in the ventricular walls of the cord thalamus of bony fishes where they posses cilia. These are sensory cells.

The choroid plexus, consisting of layers of mesodermal tissues, is a highly vascularized area in holosteans and elasmobranches. This is also known as leptomeninx. Choroid plexi or tala choroidea are filled with cerebrospinal fluid. They have many blood vessels. Its function is to provide nutrients to the nervous tissue.

4.1.4. The Brain

The brain (Fig-4.2) can be considered as an anterior enlargement of the spinal cord with hypertrophied centers associated with the development of inputs from the special sense organs. The neural tube formed by inrolling and fusion becomes progressively sub-divided anteriorly by constrictions as development proceeds, so that there are swellings along the anterior end of the tube, the neuromeres separated by these constrictions. The three anterior most neuromeres are larger than the others, which forms the two parts i.e. telencephalon (cerebral hemispheres) and diencephalon, both in the forebrain (prosencephalon) and midbrain (mesencephalon).

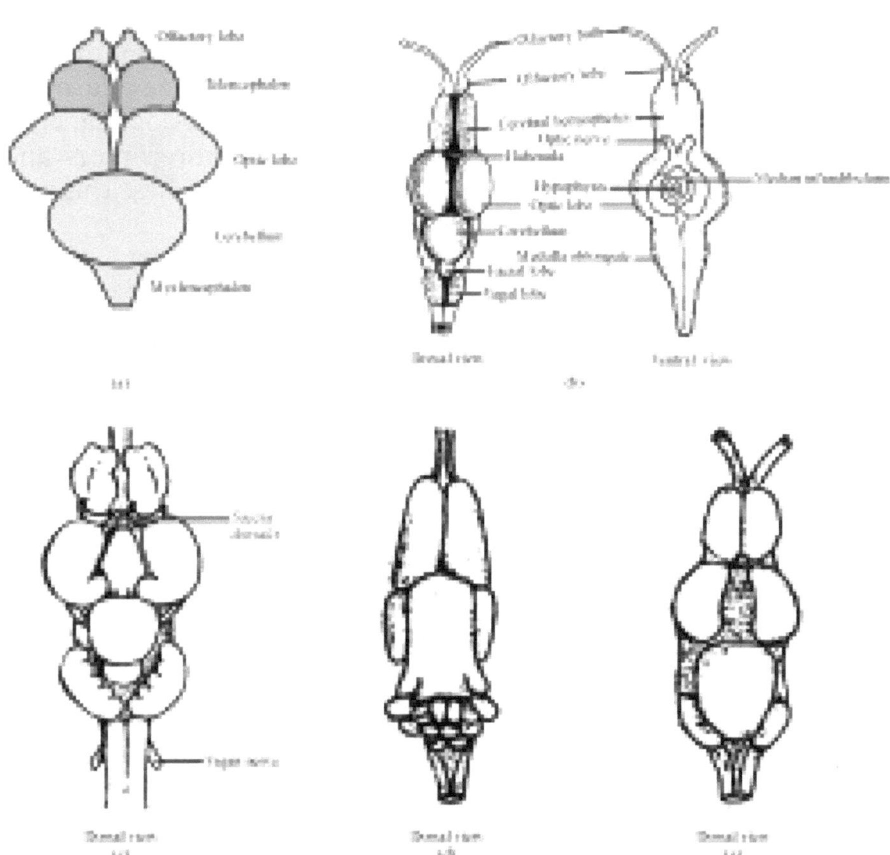

Fig. 4.2 : Brain of rainbow trout (a) *Puntius ticto* (b), *Cyprinus* (c), *Tor tor* (d) and *Mystus seenghala* (e).

Behind these, seven more neuromeres subdivide the major parts of the hindbrain (rhombencephalon).The medulla oblongata elongates to form spinal cord, where the narrow central canal is the posterior continuation of the brain ventricular system. Clusters of primary neurons are found near the center of each neuromere. The boundaries of the neuromeres coincide with regions where specific regulatory homebox genes are expressed early in brain development. The absolute size of brain differs in many fish during its ontogeny but the relative size of different brain regions, the ratio between total adult brain weight and body weight is similar for all fish species (Fig-4.2) (except elasmobranches). Fish can be considered as animals with little brain compared to higher vertebrates. In birds and mammals, the brain: body weight ratios are as much as 400% greater than fishes. A number of factors influence brain size, such as neuron number and complexity (i.e. size of dendritic fold) and changes in the complexity of the circuitial linking of neurons are comparable with that of mammals. Elasmobranchs have large size brain and their brain size and many other features like elaborate mesencephalic system of dopaminergic neurons. Elasmobranches have an elaborate electroreceptors and proprioreceptor system, which are not found in other fishes. This may be the reason for their large brain size.

4.1.4.1. Parts of the brain

A. Forebrain (prosencephalon) - Prior to telencephalon, the anterior most part of the brain, there is a pair of solid olfactory lobes, which receive and process signals from the nostril via the two olfactory nerves (Fig. 4.2). In fishes, the olfactory lobes are very large as they hunt primarily by smell. The olfactory nerves originate from these lobes and end in front in a pair of bulbs at the base of olfactory rosesette. In some species like *Tor tor, Mystus seenghala* only the bulbs are present at the fore ends of the olfactory nervers.In contrast, *Channa striatus* has olfactory lobes instead of bulbs.

Behind the olfactory lobes, the bilobed telencephalon, equivalent to the solid mass of cerebrum or cerebral hemispheres of higher vertebrates, is present. A thin non-nervous membranous pallium and sub-pallium cover the roof of the diencephalon (Nieuwenhuys, 1966). The small space between the thin roof and the hemisphere are considered as brain ventricles I and II. A bundle of nerves forms the anterior commissure that joins telencephalon with diencephalon. Many other nerve tracts are also connected to this commissure nerve. Besides the reception of smell, telencephalon is believed to play important role in reproductive behavior (Demaski *et al.,* 1975). The telencephalon also facilitates activities of lower brain centers and mechanism (Aronsen, 1970).

The diencephalon is covered by the posterior bulging of the cerebral hemispheres. It is divided into dorsal epithalamus consisting the epiphysis or pineal organ (Shrivastava *et al.,* 1980) and two equal size habenular ganglia, the thalamus and hypothalamus. The nerve fibers from telencephalon connect the thalamus, hypothalamus and the olfactory lobe. Pineal is sensitive to photoreceptor. It acts as a baro or chemoreceptor for cerebrospinal fluid or helps in facilitating the olfactory responses to sex hormones. The thalamus receives both direct retinal and secondary visual input from the optic tectum, descending telecephalon fibers. The ventral part of thalamus possesses geniculate nuclei or ganglia (Singh, 1971). This ganglion is well developed in sharks and is known as geniculate lobes. The hypothalamus receives nerve fibers from the telencephalon and taste (gustatory) fibers from medulla and sends efferent fibers to the reticular centers of the brainstem. There are two nuclei centers present in the hypothalamus - the nucleus preopticus or supraoptic nucleus and nucleus lateralis tuberis. Both are made up of secretory nerve cells. The large neurosecretory cells of supraoptic nucleus send granular neurosecretory materials to the underlying neural lobe of the endocrine gland, the pituitary or hypophysis. Sacculus vasculosus is present as pigments folded sac at the hind end of the hypothalamus whose function is not yet known, and may perhaps concerned with secretion and resorption of ventricular fluid. Hypothalamus regulates feeding, escape, attack, sexual behavior, and homeostatic control of body functions. A pouch like down growth projects from the ventral side of the hypothalamus, which is known as infundibulum and at its tip the hypophysis is attached by a stalk. The infundibulum is enlarged laterally to form bean-shaped inferior lobes, which remains ventral and opposite to the optic lobes. A pair of optic nerves that cross each other to form optic chaisma, enter the brain in the antero-dorsal part of the inferior lobes. Diencephalon is an important correlation center for afferent and efferent impulses. Hypothalamus influences the endocrine system through hypophysis.

B. Mid Brain (Mesencephalon) - It is relatively large. It has a dorsal optic tectum and a ventral tegmentum. The optic tectum has two optic lobes with numbers of zones of different sized and shaped neurons (Khanna and Singh, 1966). The optic tactum is composed of six layers. In higher bony fishes, the optic tectum projects into the optocoel in the form of torus longitudinalis, which connects with the posterior commissure. The fibers of the optic nerves end in the tactum and the image formed on the retina is projected on it. Like other vertebrates, fish have convex lenses in their eyes that create an inverted image on the retina through the special tectal nerve arrangement. The optic tactum is comparable with the cerebral cortex of mammals in its structure and function. This part of the brain is the co-ordinating center for the eye-body reactions besides doing other functions of the brain such as learning, responses to smell etc. The optic tactum differs in its degree of development in various fish species. It s reduced in blind and cave fishes.

C. Hind Brain (Rhombencephalon) - Cerebellum or metencephalon develops as a distinct out growth from the upper rim of the 4th ventricle of brain, medulla (myelencephalon). In higher bony fishes, its anterior part enters into the cavity of the optic lobes as valvular cerebelli, and so it is partially visible as corpus cerebelli. The corpus cerebelli has an outer molecular and inner granular layer and vice versa in valvular cerebelli. The molecular or chief receptive cell layer and Purkanje cells assume prominence in the cerebellum of higher vertebrates. In lampreys, cerebellum appears only as small dorsal bridge between acoustic-lateral line nerve areas. In hagfish it is represented by a small commissure that consists of eight nerves and lateral line fibers forming octavo-lateralis system. In elasmobranches , the cerebellum size varies with the size of fish i.e. small sharks have simple bilobed cerebellum but in predatory sharks (*Lamnidae*) it is the largest statics part of the brain. Distinct lateral and auricular outgrowth may occur from cerebellum that is connected to lateral line and vestibular sense organs. Enlarged cerebellum is presumed to be involved in reception of electrical impulses (Khanna and Singh, 1967). The cavity (or ventricle) of cerebellum, known as metacoel, is prominent in elasmobranches but completely disappeared in higher bony fishes. The main functions of cerebellum are to control swimming equilibrium, maintenance and co-ordination of muscular tonus and orientation in space (Singh, 1993).

Medulla oblongata (brain steam) or myelencephalon is the last part of the brain from where the ganglia or nuclei are originated and the endings of sensory input of all the ten pairs of cranial nerves, except the olfactory (I) and optic (II) nerves occur. The boundary between medulla oblongata and spinal cord of fish is indistinct. The walls of medulla are thickened by the primary nuclei of the acoustico - lateralis system and the basal central core contains the reticular formation. This formation is important in elasmobranches since it contains very large neurons with extensive dendritic arborisations. The medulla can be divided into columns of nerve fibers based on the type of information transmitted. Medulla contains a cavity inside, known as 4th ventricle. Fishes like *Mugil cephalus, Clupea pallasi* possess paired swellings called "cristae cerebelli" on the anterolateral boundary of the 4th ventricle. These cristae are associated with schooling behavior of the fishes. Nerve connections to the cristae are not known in fishes. Other fishes like goldfish (*Carassius*), carpsucker (*Carpiodes*) have vegal lobes posterior to the cristae from which cranial nerve IX and X emerge. Such fishes also have a palatal organ in the roof of the mouth where food is tasted by taste and touch. In tuberculum imper or facial lobe, a central hillock posterior to cerebellum is noticed in cypriniforms. This lobe arises from a fusion of root elements of the facial (VII) and vegal (X) nerves, where gustatory and tactile impulses are correlated with visceral sensory nerves as in carp (*Cyprinus carpio*). Species having a better hearing sense possesses a well formed central accoustic lobe.

Cranial nerves III to X arise from this portion of the brainstem. The afferent nerve fibers of the medulla can be divided into somatic and special sensory nervous inflows with cranial nerve VII (facial), VIII (stato-acoustical), and IX (glossopharyngeal) and X (vagus) (mostly sensory). There are more somatic sensory fibers in lamprey than elasmobranches. But these nerve fibers are further reduced in bony fishes (Osteichthys).

4.1.5. The Spinal Cord

The circular spinal cord originates from the brain and extends for whole length of the body inside the vertebral canal. It is formed by the inrolling and fusion of neural folds. In Cyclostomes, it is flattened due to the absence of intra-medullary blood supply within the spinal cord. In Condrichthys, it is comparatively more differentiated. It has cruciform grey matter of nerve cells with a central canal and the grey matter is surrounded by the white fiber tracts in the peripheral area (Fig- 4.3). The grey and white color is based on the staining properties of the myelin sheaths of the nerve fiber with iron hematoxylin and also due to the presence of dense nerve cells in grey matter. The central canal is filled with cerebrospinal fluid secreted by brain. The cord consists of series of segments from which the dorsal and ventral roof arise and these two roots join to form spinal nerves (Bone, 1963). At the larval stage of many fish species large Rohon-Beard cells are present dorsally in spinal cord, which mostly disappears in the adult stage of advanced fishes, (Bone, 1963).

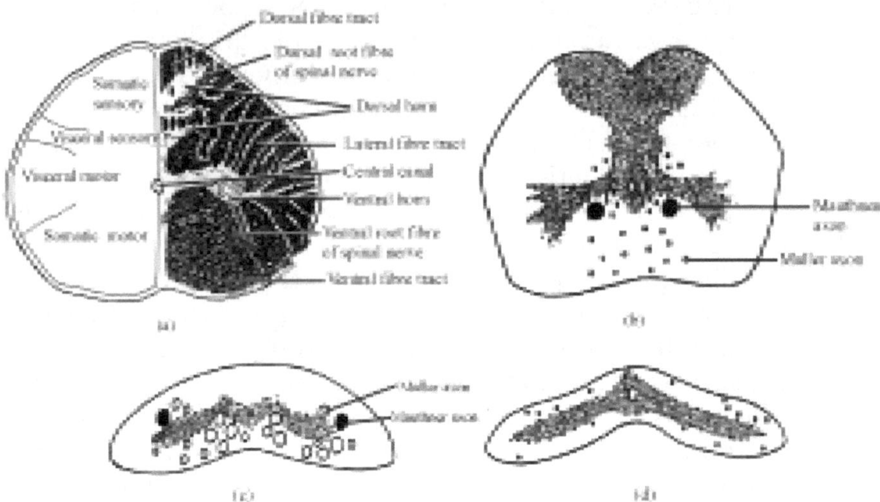

Fig. 4.3 : Schematic diagram showing the functional organization in the cross section of spinal cord of shark (a), teleost (b), lamprey (c) and hagfish (d). (Modified from Ravainen *et al.*, 1973. J. Comp. Neurol. 149:193.)

In Chondrichthys and Osteicthys, the grey matter has the paired dorsal and ventral horns (Fig-4.3). The dorsal horns receive the somatic and visceral sensory fibers and the ventral horns contain motor fibers. The neurons possess extensive dendritic fields and individual neuron sends dendrites from grey into white matter. Commissural cell axons and other nerve fibers are more in ventral horn. In Cyclostomes, the neural dendrites extend from the ventral grey column into dorsal white matter. Commissural cells are found in the cord. The axons of these cells extend ventral to the central canal and then bifurcate to ascend or descend within the spinal cord. Mullar cells constitute a conspicuous system of descending motor co-ordinating nerve fibers.

In bony fishes, the grey matter has many large motor horn cells in groups. The cell groups are located in the dorsomedial or ventral grey and both are efferent. One is efferent to tail musculature and the other group is efferent specialized area like pectoral fin. Commissural neurons and nerve fibers are also found in these fishes. Many nerve fibers are connected to the propiospinal system and may ascend to the medulla, cerebellum or may be to the roof of midbrain (Nieuwenhuys, 1964).

4.1.6. The Nerves

4.1.6.1. Spinal nerves

Segmentally arranged dorsal and ventral roots emerge separately from the spinal cord in all fishes, except lampreys (the two nerves do not unite) and then join to form the mixed sensory and motor spinal nerves. In hagfish, they occur as true mixed spinal nerves. The jawed fishes have dorsal root ganglia for the neurons of the sensory (dorsal) root nerves that send their sensory fibers to the periphery. The ventral root contains mainly the axons of the spinal somatomotor neurons, which pass to the myotomal and fin musculature, and in addition, visceromotor autonomic fibers. In the adult dogfish (*Scyliorhinus*), the ventral root of the abdominal region contains about 350 fibers but in zebra fish (*Brachydanio*) each ventral root has nearly 100 somatomotor axons. The ventral root passes out of the spinal cord between the myotomes. Thus a single ventral root contains the axons of motor neurons innervating two adjacent myotomes.

In some rays (Triglidae) the long and separate anterior rays of the pectoral fins carry special receptor for tectile and chemical clue. Sensory nerves from these rays are marked on the spinal cord by separate swellings corresponding to their dorsal root nerves and their fiber projections also reach the medulla and the midbrain.

Swimming in fish is due to endogenous rhythm from the posterior medulla. This, in turn, passes on to the cord and is influenced at all levels by outside stimuli. There is no autonomic swimming activity regulating center but that swimming movements are of reflex nature and depend on the coordination of afferent impulses on spinal cord and medulla. Swimming is produced by descending glutaminergic pathways synapsing with the local spinal segmental neuron pattern.

4.1.6.2. Cranial nerves

Cranial nerves are the peripheral nerves that arise from the brain as distinct from spinal nerves. There are ten pairs (I to X) of cranial nerves (Fig-4.4), out of which some may consist of sensory fibers only, some motor fibers only and others contain both types of fibers. Olfactory and optic nerves consists special sensory nerves fibers, oculomotor and trochlear have motor fibers and abduceans, facial, glossopharyngeal and vagus consist of mixed fibers. Cranial nerves do not have dorsal and ventral roots as in case of spinal nerves.

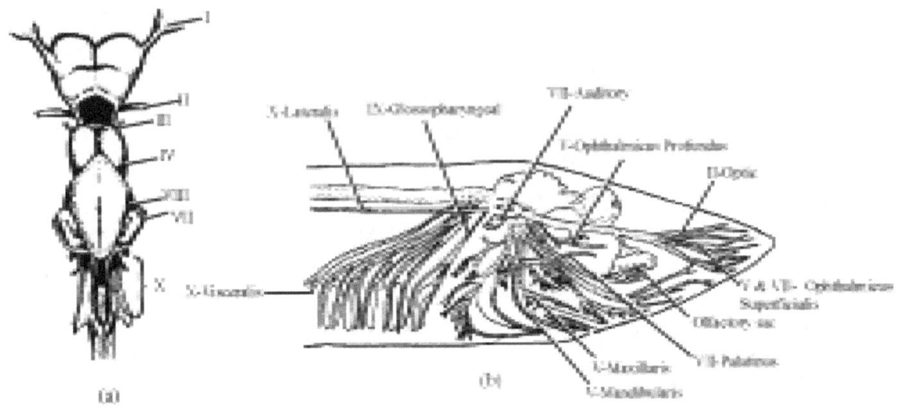

Fig. 4.4 : Schematic diagram showing the origin of cranial nerves (I to X) from brain of shark (a) and the different cranial nerves (b).

The ten cranial nerves are as follows :

1. Olfactory (I) - It arises from the olfactory lobe, particularly lamina terminalis of forebrain. It extends into snout to end in olfactory rosette. It is a special sensory nerve. It carries small impulses to brain.
2. Optic (II) - It starts from the ventral side of the optic tectum of midbrain. It is a sensory nerve and supplies nerves to the retina by entering through the optic foramina. It carries visual impulses to brain.
3. Oculomotor (III) - Arising from the ventral side of optic tectum, it supplies superior medial inferior rectus and inferior oblique muscles of the eye passing through the optic foramen. It is somatic motor and innervates four to six striated eye muscles.
4. Trochlear (IV) - This nerve is also involved with the movement of eye ball. It arises from dorsolateral surface of the hinder margin of the cerebellum. It enters into the superior oblique muscle of eye ball, one of the six striated muscle of the eye.
5. Trigiminal (V) - It is one of the largest nerves that arise from the anterio-lateral side of medulla. It is mixed somatic sensory and motor in function. It bears a large semilunar or Gasserian ganglion near its origin. It has three main trunks; one each enters into ophthalmic, maxillary and mandibullar. These three trunks arise just behind the ganglion. The ophthalmic trunk has two branches such as ophthalmicus superficialis and ophthalmicus profundus. Both of these are sensory.

Maxillary trunk runs along the margin of upper jaw and mandibular goes to the lower jaw.

6. Abduscence (VI) - This nerve is also for eye ball. It starts from the ventral surface of medulla and runs behind the trigiminal to supply to the lateral rectus muscle of eye.

7. Facial (VII) - It originate from the sides of medulla behind the Vth nerve. It quickly joins the Vth nerve to form trigeminofacial complex. It is a mixed nerve. It divides into supra-orbital, infra-orbital and hyomandibular branches. The supra-orbital further branches into ophthalmicus superficialis trigeminalis and ophthalmicus superficialis facialis.These two branches innervate lateral line system. The infra-orbital runs ventral to supra-orbital and divides into four branches such as maxillaries, buccalis, mandibularis and palatine. The hyomandibular runs just behind the mandibular nerve to supply the lower jaw.

8. Auditory (Accoustic) (VIII) - Due to the common ganglia, this nerve is related to facial nerve. It originates from the side of medulla and runs behind the facial nerve. It bifurcates into two branches. The vestibular branch goes to utricular and ampullae of the internal ear and the other saccular (or cochlear) supplies to sacculus and lagna. Both are sensory nerves.

9. Glossopharyngeal (IX) - It develops from the ventero-lateral side of the medulla, immediately behind the auditory nerve and it is close to vagus (X) nerve. It is a mixed nerve but more like the spinal nerves. It enters into the first gill cleft and divides above it into pre—and post-trematic rami. It innervates pharyngeal mucosa and supplies the hyoid arch, oral cavity and taste bud. It is both sensory and motor.

10. Vagus (X) - It is a mixed nerve. It originates from medulla just after the IXth nerve. Before supplying to different body organs through its branches, it divides into five branches such as:

    a. Supratemporal branch - supply to skin, lateral line of supratemporal region.

    b. Dorsal recurrent branch - supply to taste bud.

    c. Lateral line branch - innervates lateral line.

d. Visceral branch - goes into visceral organ.

e. Branchial branch - supply and receive nerves from four posterior gills.

Optic nerve (II) is not a part of segmental arrangement (dorsal and ventral root). It arises as out-pouching of the brain. So, this nerve is not comparable to the cranial nerves (Bone *et al.*, 1995). Terminalis nerve (O) is a small bundle of fibers entering the brain ventrally just behind the olfactory nerve.

4.1.7. Autonomic Nervous System

This nerve system is concerned with the internal body environment of fish. It is divided into sympathetic and parasympathetic. This system consists of ganglia and nerve fibers of muscles, glands and other body vital organs, and thus controls the body metabolism. The important features of this nervous system is that the nerve fibers (preganglionic) after leaving the CNS, form synapses with a second group of nerve fibers (postganglionic). In lampreys and hagfish, both the heart and viscera are innervated by vagus nerves. Subcutaneous nerve plexi, that are innervated by spinal nerves are of sympathetic system and are found more at the orifices of the slime glands of hagfish. In elasmobranches, there is no sympathetic system in head. The irregular sympathetic ganglia are found along the side of the vertebral column. There is at least one ganglion in each segment. The ganglia are situated in the dorsal wall of the posterior cardinal sinus in close association with chromaffin tissue. The smooth muscle of the gut and anterior walls receive the sympathetic nerve. The posterior ganglia innervate the genital duct, kidneys, urinary bladder and rectum. In sharks, the heart and organs of digestive tract are supplied by vagus nerves. In bony fishes, a sympathetic chain is formed up to cranial nerve V by the joining of segmentally arranged ganglia. Most preganglionic fibers for sympathetic ganglia of the head originate from the tail end of spinal cord. In teleosts sympathetic connections to skin nerves are noticed. Certain postganglionic sympathetic fibers bring change in color of teleosts. The sympathetic cardio-accelerating fibers of higher vertebrates is lacking in fish.

There is apparent lacking of parasympathetic system in fishes and this system is almost exclusively represented by a part of wildly branched vagus (cranial nerve X). Cranial nerve III innervates the eyes. Vagal stimulation has a inhibitory action on the sinus venosus and the auricle but without any effect on ventricle.

4.1.8. Neural Physiology

The properties of the ion channels of the neuron membranes are responsible for their electrical activity. This ion channel is the Na^+ - K^+ channel or pump. This channel is modulated by neurotransmitters or by hormones. It causes behavioral changes. Modulator neurons with endogenous beating activity and wide projections are found in teleosts. Such neurons with endogenous rhythmic activity are also found in higher vertebrates that contains serotonin (5-hydroxytryptamine), noradrenaline, dopamine and histamine. In fishes, these neurons form a small ventral group at the entry of nerve 0 to the telencephalon. Except cerebellum and spinal code, these neurons enter into all parts of the brain. They contain gonadotropin releasing hormone as the neuromodulators. Under the influence of the environmental stimuli or hormones, these neurons change their rhythmic action which in turn has an effect on central neural activity.

Most sensory mechanisms in fish work with full efficiency at the normal level of stimulation. Electrical shock conditioning to visual and olfactory stimuli lead to learning accomplishments (Laglar *et al.*,1977). The accomplishments are detected by altered heart beat (an autonomic response) or by swimming movements (a motor response). Fishes are capable of reflex learning. A second order conditional reflexes are also established by fishes. After training, the goldfish is able to transfer a response learned on the basis of olfactory signals to visual clues. It can also make a similar transfer from auditory to thermal clues. Retention of learned responses with a long lasting memory is noticed in migratory fishes like salmons (*Oncorhynchus and Salmon*). They retain steam odor, learned during juvenile period throughout their entire lives (2-6 years). The home steam odor memory is acquired by an imprinting process. This helps them to return home after migration. In many teleosts, electrical stimulation of a localized region in the midbrain tegmentum excites coordinated swimming movements. Similarly electrical stimulation of area near the lateral recess of third ventricle in the inferior lobe of the hypothalamus causes low-threshold and complete feeding responses in fishes. Similar centers may well exist for reproductive behavior. The basis of behavior components are patterns of neuromuscular co-ordination that lead to movements in the wake of external and internal stimuli. These patterns are genetically determined and the movements that make up the behavior train during reproductive activities recur similarly in all individual species. Fishes perform species-typical movements of courtship and/or nest-building. Therefore the pattern and sequence of these activities is a part of genetic makeup of the fish as is its body shape or coloration. Interplay between genetic and environmental factors during evolution has designed the pattern and the responses of the nervous system, including the receptors.

In teleosts, the tectum is multilayered with retinotopic maps of the visual field and a complex set of neurons having long dendrites for the light stimuli from the responses of visual cells. The receptive fields of the cells differ. Those with well-defined boundaries are probably involved in dealing with location and colour of objects, whilst those with irregular and poorly-defined boundaries respond to brightness, contrast and velocity. There are neurons that are sensitive to light in CNS, as observed in minnows (*Phoxinus*) who can be trained to respond to changes of illumination after removal of eyes (Bone *et al.*, 1995). In lampreys, light-sensitive neurons occur in the caudal spinal cord.

In electroreceptive fishes, the lateral line lobes of the hindbrain have a special dorsal electroreceptor nucleus, which projects to the cerebellum. This system is used by the fish to detect the prey. The cell arrangement and circuitry of the electric lobes of the cerebellum is similar to that of non-electric fishes, but the dendrites of the Purkinje cells are arranged in an extra-ordinarily regular manner between sheets of parallel fibers, in much more ordered arrangement. The Purkinje axons end on enormous eurydendroid cells and these axons are the excitatory output from the electric lobes.

4.2. NERVOUS SYSTEM OF SHELLFISH

In the basic arrangement, the nervous system of arthropods is similar to annelids. The main differences in this system of the two animal groups are:

1. Great development of the ganglia, particularly the cerebral and sub-esophageal.
2. The reduction in number of the ganglia corresponding to the reduction in number of the body segments.
3. The extension of one neuron over several segments.

The gross morphology of the CNS varies with that of the body of arthropods, four pairs of nerves from these are larger than the others. These four pairs are :

1. The nerves to the antennules.
2. The optic nerves from eye.
3. Tegumentary nerves from epidermal sense organs, with small branches to the excretory bladder.

The nervous system of decapods consists of CNS linked to peripheral neurons, the CNS taking the form of a nerve cord ventral to the gut and the supra-esophageal ganglia or cerebral ganglia joined to the nerve cord by two long commissures surrounding the esophagus.

4.2.1. The Brain (Supra-esophageal Ganglia or Cerebral Ganglia)

The brain or cerebral ganglia of arthropods are considerably larger than annelids, which is correlated with well-developed sense organs. Many arthropod groups show complex behavioural patterns. The nerves originating from brain run to the antennules, the antennae, the compound eye and statocytes. This ganglia are in reality double structures, being composed of the *archicerebrum* (the ganglia of the pre-antennary segments) and the *syn-cerebrum* (the fused ganglia of the antennulary and antennary segments). The head of all arthropods contain two or three preoral segments and the antennae are segmental appendages. These anterior ganglia of the head are fused to form the brain. The brain consists of three major parts (Fig-4.5):

1. An anterior protocerebrum.
2. A middle deutocerebrum.
3. A posterior tritocerebrum.

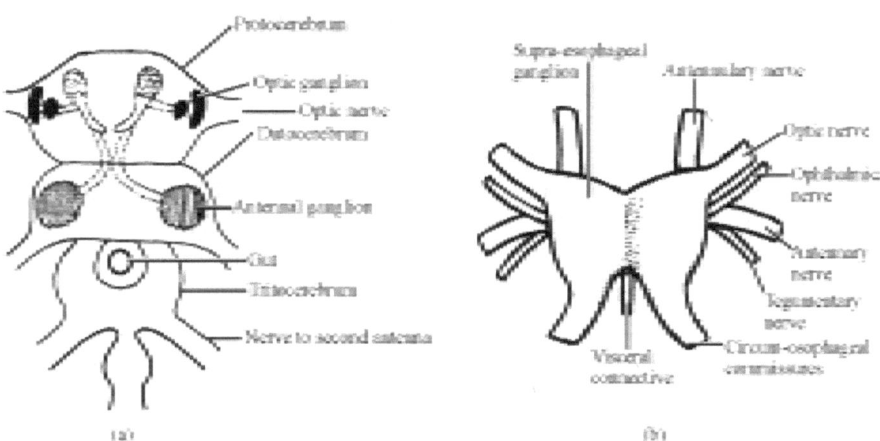

Fig. 4.5 : Different regions of the mandibulate arthropod brain (a) and shrimp brain with the origin of nerves (b).

The protocerebrum contains pairs of optic centers (neuropiles) through which nerves from eyes enter. The optic and other neuropiles of the protocerebrum function in integrating photoreceptor, body movements and to initiate the complex behaviour. The deutocerebrum receives the antennal nerves and contains their association centers. The cheelicerates lack antennae, and so, the deutocerebrum is absent in their brain. The nerves from tritocerebrum innervate the lower lip (labium), the digestive tract (stomatogastric nerves), the cheelicerae of cheelicerates, and the second antennae of crustaceans. With regard to the origins of the protocerebrum and deutocerebrum, there are two general opinions. One opinion holds that these two brain regions represent the non-segmented acorn or archicerebrum, which is similar to the cerebral ganglia of many annelids.

The general plan of the nervous system in crabs is the same as in shrimps or other decapods (Fig-4.6). One great difference is the fusion of all the ventral ganglia in the cephalothorax into a single massive supraesophageal ganglion in crabs (Fig-4.6b). The supraesophageal ganglion lies immediately behind the rostrum. Nerves that can be seen connecting on each side with supraesophageal ganglion are the antennulary, optic, oculomotor, tegumentary and antennary. Posteriorly there are two fine nerves that connect with the stomatogastric system.

4.2.2. Ventral Thoracic Ganglionic Mass

The crowding together of the appendages behind the mouth in prawns is reflected by fusion of the first five pairs of ganglia behind the mouth to form a single subesophageal ganglia. There is a conspicuous hole near the center of the ganglia for the descending sternal artery (Fig-4.6a). The sternal artery passes between the commissures joining the ganglia in segments 12 and 13 in the thoracic region where the nerve cord lies in the sternal sinus within the endophragmal skeleton. The endophragmal skeleton is formed by refter-like ingrowths of the exoskeleton in the mid ventral region of the cephalothorax. The brain receives sensory inputs from the anterior concentration of sensory organs, and also acts in motor control of the anterior head appendages. The segmental ganglias receive sensory input from the segments they serve and provide motor nerves to the associated appendages. Two minute nerves belonging to the so-called stomatogastric system leave this ganglion, one of than passes forward and medially to meet another from the other side under the stomach.

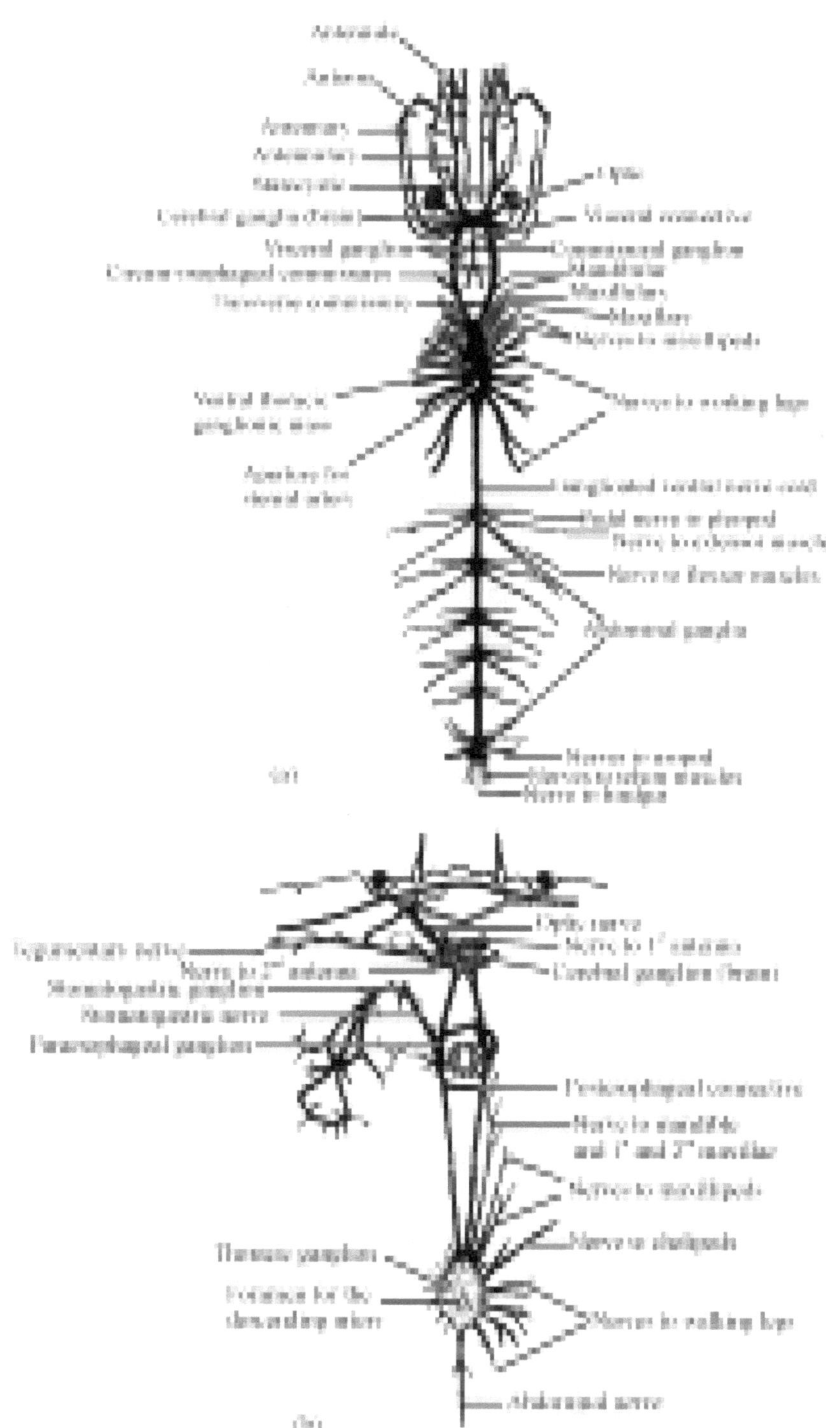

Fig. 4.6 : Nervous system of decapods - prawn (a) and crab (b).

In crabs (Fig-4.6b), ventral ganglion lies near the floor of the cephalothorax, in the fourth and fifth thoracic somites. The abdominal ganglia are reduced and fused, in association with the evolution of a compact thorax and vestigial abdomen.

The subesophageal ganglia forms the indistinguishable anterior part of the elongated ventral thoracic ganglionic mass. This ganglionic mass represents the fusion of 11 pairs of ganglia and gives off laterally 11 pairs of nerves.

4.2.3. Circum-Esophageal Commissures

Posteriorly the brain gives of a pair of strong nerves known as circum-esophageal commissures. In prawns, this pair of nerves run both forward and backward and after encircling the esophagus, unit ventrally with subesophageal ganglia (Fig-4.6a). The two commissures are crossed over, just behind the esophagus by a double bridge of tough connective tissue, called endosternite. Each commissure bears a small commissural ganglion near its anterior end, sends a small nerve to the mandible of its side and a thin transverse connective to the fellow of the opposite side near its posterior end.

In crabs (Fig- 4.6b), the circum-esophageal connectives are long over two-third of this length being behind the esophagus. The tritocerebral commissure, connective ganglion and connections with the stomatogastric system, are essentially as in the shrimps.

4.2.4. Ventral Nerve Cord

The ventral thoracic ganglion from its hind end gives rise to the ventral or abdominal nerve cord that runs along the mid-ventral line of the abdomen and in each abdominal segment it enlarges to form abdominal ganglion. Each of the first five abdominal ganglia gives of three pairs of nerves in its segments.

1. One pair of pedal nerves to the pleopods.
2. One pair of nerves to the extensor muscles.
3. One pair of nerves to the flexor muscles of the next segment. The sixth abdominal ganglion is known as stellate ganglion. This is relatively large as it is formed by the fusion of several ganglion.

4.2.5. Nervous System

Crustacean muscles contain varying properties of fast and slow muscle fibers and may be innervated by fast and slow, and inhibitory nerve fibers. Different blends of these components are used to provide the strength, speed and duration of muscle contractions needed to effect such diverse activities as rapid escape reaction (tail flip), fast or slow walking and swimming, gut contractions, maintaining posture and holding material with cheelate legs. A further layer of response sophistication is added by different types of neuromuscular synapses and neurotransmitters.

The autonomic, sympathetic or visceral nervous system consists of a few ganglia and nerves. A small nerve, originating mid-posteriorly from the brain, contains two visceral or esophageal ganglia lying one behind the other. The first ganglion is joined with the two commissural ganglia through a pair of connectives. The second ganglion gives off two pairs of nerves to the esophageal walls and cardiac stomach.

From supraesophageal ganglion a number of nerves radiate out of which four pairs are larger. Out of the eleven pairs of nerves from the ventral thoracic ganglionic mass, first three pairs are of cephalic nerves, supplying the mandibles, maxillulae and maxillae, respectively. Last eight pairs are of thoracic nerves of which the first three pairs supply the three pairs of maxillipeds and the remaining five pairs supply the five pairs of walking legs. Each nerve to a leg becomes bifurcated before entering it.

In the crabs, from the periphery of the supraesophageal ganglion nerves radiate out on each side to supply the series of appendages from the mandible to the fifth periopod. The nerves for the first five of these appendages are very small. Those for the mandible and maxillule are fused together. The nerve for the third maxilliped is somewhat larger and that for the cheeliped is largest of all. Posteriorly a small double nerve passes back in the midline into the abdomen. From the dorsal surface of the ganglion, at the same level as the nerves to the third maxillipeds, arises a pair of fairly large cutaneous nerves. Each of these nerves runs latero-anteriorly, just behind the nerve to the second maxillipeds, and then curves upwards inside the body wall, to supply the lining of the carapace and the roof of the gill.

Chapter - 5

Respiration and Gas Exchange

The respiration of animals includes the exchange of gases either at the surface of the body or through the internal processes involving circulatory system. The gases that are exchanged are oxygen (O_2), carbon dioxide (CO_2) and the inert gas like nitrogen. When the organism respires, the O_2 that is taken up from the external environment is transported by the vascular system to individual cells within the tissues. At the same time this system picks up CO_2, an end product of cellular metabolism, and releases it to the environment. Thus respiration is a term applicable either to a whole organism or one of its cells. The mechanical process, by which respiration is achieved, is called ventilation. This ventilation rate of aquatic organisms is controlled directly by the concentration of dissolve oxygen (DO_2) in water.

Respiration in aquatic medium is produced by a water current passing in a single direction. The area of respiratory surface is thus an important limiting factor in the movement and growth of the animal. For aquatic animals, if the water is at all deficient in O_2, the rate of breathing is naturally accelerated, and the animal appears to respire hurriedly. Other factors, which lead to an increased rate of respiration, are the stresses.

Oxygen as a gas has a low solubility in water. The amount of O_2 content of water varies with the temperature and salinity in a predictable manner. Less O_2 can be held in fully air-saturated warm seawater than fully air-saturated cold freshwater. At the surface, air saturated water at 20°C contains about 3% of O_2; while the O_2 content of the water sets the absolute availability of O_2 in water. It is the O_2 partial pressure (PO_2) gradient that determines how rapidly O_2 can move from water into the aquatic animal's blood to support its aerobic metabolic rate. This is because O_2 moves by diffusion across the gill of the animal. Since O_2 can cross cell surfaces by diffusion, the gas exchanger to acquire O_2 from the water must have a large area. Most fish, mollusks and crustaceans have gills in internal chamber through which water is pumped to increase still further efficiency of exchange. Since water is dense and viscous, a relatively large expenditure of energy is needed to force water to flow around the gas exchanger. The flow of water over the gills is always unidirectional, whether it is pumped or simply flows over the gill and the blood flow in the gill can be arranged in the opposite direction to create a counter-current effect. If the blood flows in the opposite direction, the maximum PO_2 of the blood leaving the exchanger could be very close to that of the incoming (ambient) water and above that of the exhaled water.

5.1. RESPIRATION AND GAS EXCHANGE IN FISH

In fishes, the exchange of gases takes place through the integument, gut, gills, lungs and air sacs. In tiny fish larvae, the integument itself is adequate for respiration. But at adult stage they respire through gills. Some fishes have auxiliary systems for exchanging gases directly with the air either when the water is low in O_2 or when the water recedes during either tidal or annual cycles. Oxygen consumption varies greatly with species, size, activity, season and temperature. Large animals use less O_2 per unit weight than small animals do. Consumption increases with temperature up to a critical level then falls off. Fast- swimming fishes use much more O_2 during active periods than sluggish ones. An example of a marine species is mackerel (*Rostralliger kanagurta*). It has large gill surface and its blood will carry more oxygen then other fishes. As it needs to swim almost continuously, its gills are ventilated adequately even through the water in which it lives is always almost completely saturated with O_2. Freshwater species with high O_2 requirements include the trouts (*Salmo gairdinari*). These fish live habitually in the oxygen-saturated water of cold streams. They cannot long withstand water with less than 5 mg/ liter of O_2. On the other hand, freshwater fishes such as carp and catfish can survive with DO_2 of only 0.5 mg / liter.

5.1.1. Gill

The egg of elasmobranches and teleosts are large in size, and the larvae hatch with functional gills. But when the hatchling is very small without any functional gill, cutaneous respiration across the body surface becomes the process of respiration. As the larvae have less complex organ systems and surface-to-volume ratio is more, cutaneous respiration is sufficient at that stage. However, with the increase in size of the larvae, the surface-to-volume ratio decreases, and cutaneous respiration becomes insufficient. So, the functional gills develop. Hagfish and some teleosts also depend on cutaneous respiration at adult stage.

The lateral walls of the pharynx are perforated by means of a series of slit-like apertures to form visceral clefts. These clefts become modified for respiratory purpose. The first cleft, called spiracle, remains between the mandibular and hyoid arch. The second or the hyoidean cleft lies between the hyoid arch and the first branchial arch and the rest of the clefts are situated between the succeeding branchial arches. Each gill cleft communicates with the pharynx and the exterior by an internal branchial aperture and an external branchial aperture, respectively. Teleosts have a single branchial aperture on each side of the head due to the development of an operculum covering the gills. The gill clefts bear internally a highly vascularised plate-like structure called gills or branches (Fig-5.1). The clefts are separated from each other internally by interbranchial septa, which are covered externally and internally by mesodermal membrane and endodermal epithelium of pharynx, respectively. Each septum is supported by cartilage or bone.

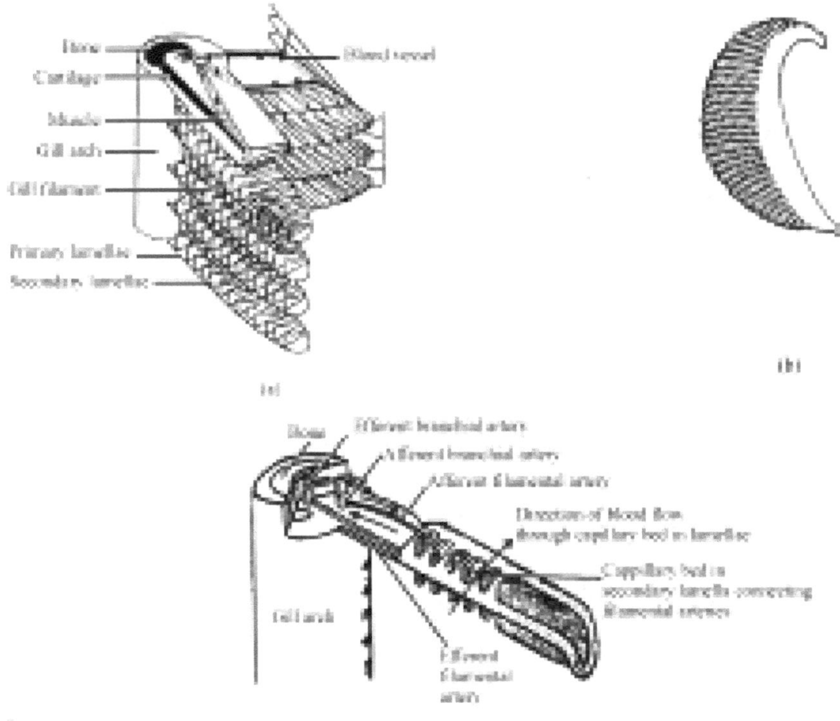

Fig. 5.1 : Diagram of fish gill structure (a), single hemibranch (b) and blood flow in gill filament (c)

The gills are of three types according to the arrangement of the gill lamellae (Fig- 5.1). The gill types are :

1. Hemibranch or demibranch or half gill (Fig-5.1b) where a series of lamellae are present on one side of an interbranchial septum.
2. Holobranch or complete gill where two hemibranchs enclosing between them interbranchial septum make up a complete gill.
3. Pseudobranch or opercular gill, which is usually covered by a layer of mucus membrane. A hyoidean pseudobranch consisting of a series of gill filaments is present anterior to the first gill in actinopterygians like that of catla (*Catla catla*). The pseudobranch has respiratory function at the embryonic stage.

5.1.1.1. Gill design

The gills are ciliated in tunicates and amphioxus. The endostyle of the gills produce mucous to trap the food. The beating pattern of cilia allows the flow of water over the gill for gas exchange and to collect the food. Blood flowing through them is mostly deoxygenated since the ciliary tracts of the gill bars use more O_2. There is a change from the ciliary to muscular respiratory gill in fish. The muscular gill has the advantage that using the muscle the fish can change the gill angle for better unidirectional flow of water over the gill. So, muscular gills are more efficient respiratory organs. There are two types of muscles, abductor and adductor (constrictor branchialis) for the movement of the gill filaments, which are associated with the gill rays, the gill arch and the gill septum. The abductor and adductor muscles are striated or smooth, respectively. These muscles cross connect the gill rays of the oral filament with that of the aboral filament.

In hagfish, water is drawn in through the nostrils by the action of a muscular membrane known as velum and exists from a series of gill pouches via single external opening. The unidirectional water flow through the muscular gill sacs or branchial pouches is done by rolling and unrolling of velvar folds. These remain in a chamber produced from the naso-hypophyseal tract and are operated by complex muscles inserting into cartilages of the neurocranium. Contractions of gill pouches and their ducts assists in producing the flow.

The ammocoete larva of lamprey has a series of finely divided gill slits, which are ventilated by the velum. There are seven paired gill sacs in lamprey (*Petromyzon marinus*) as representative gills, which are separated from each other by a thin diaphragm. The gill sacs open towards the lumen of digestive tract. The inside of each gill sac is covered by gill filaments with secondary cross folds on them to increase the respiratory area. The sacs, supported by cartilaginous branchial arch, communicate with the exterior through epithelium-lined branchial atria which opens to the exterior via gill pores. Between each branchial atrium and its external gill pore there is a short, posteriorly directed branchial canal. Between the gill pouches, there are septa that contain venous blood sinuses, cartilaginous support from branchial arch and muscle. These interbranchial septa receive an afferent artery each from truncus arteriosus . This artery divides into anterior and posterior branchial pouch arteries and further into arteries of the filaments and capillaries for gas exchange at the lacunae in the gill lamella.

A detailed structure of the gill of teleosts has been described by Dutta Munshi (1960). Gill structure is almost similar in all Gnathostome fishes (Fig- 5.1). Elasmobranch gills differ from teleosts gills in the way that the gill filaments remain attached along their entire length (so their name is elasmobranch), which is not in teleosts. The teleosts usually have four pairs of branchial gill arches or gill bars, bearing the gills but in elasmobranchs, the hyoid arch bears a posterior respiratory hemibranch. So, there are five pairs of gill bearing arches in sharks. In some sharks (*Heptranchias*) there is up to seven gill arches (Dutta Munshi, 1960). The arches may be bony or cartilaginous. The inner border of the gill arch extends to form gill rakers. The rakers are small pointed projections guarding the gill slits. Each gill arch has at least one set each of abductor and adductor muscles. The abductor muscles are present on the outside of the gill arch connecting it with the proximal ends of the gill rays. The adductor muscles are present in the inter-branchial septum. The adductor muscles tend to bring the parts closer and reverse by the abductor muscles, causing contraction (positive pressure) or expansion (negative pressure) of the cavity.

The four hypobranchials of each side are connected ventrally to a median copula. The series have regularly-spaced folds form hemibranchs and the folds are called primary gill filament or lamellae. Two rows of primary gill filaments are held by the ceratobranchial or epibranchial segments of each gill bar. The two rows of primary gill filaments are separated by a short fleshy interbranchial septum so that the two rows are free at their distal ends. It is much shorter in *Rita rita* and *Channa striatus* but little longer in rohu (*Labeo rohita*) and *Hilsa ilisha*. Filaments of one hemibranch alternate with the filaments of the other himibranch of the same holobranch. The four holobranchs on each side, virtually separate the opercular cavity from the buccal cavity with hemibranchs projecting into the former and remaining in turn protected from any risk of abrasion by the operculum, yet allowing flow of water through them, serving like a filler.

The numerous, minute lamellae which protrude from both sides (oral and aboral) of each filament are the primary sites of gas exchange (Mayle and Cech, 1988). These are called secondary gill lamellae. The arrangement of opposing lamellae on the adjacent filament is such as to produce an effective gill sieve with minute pores. These secondary gill lamellae are set parallel to the water flow through the branchial chamber. The lamellae are made up of thin double layered epithelial cells on the outside and thin basement membrane and supportive pillar or pilaster cells on the inside (forming blood channels) allowing blood flow through the interior without significant changes in shape (Hughes and Grimstone, 1965). Amoebocytes and lymphocytes are commonly found in the epithelium of *Anabas* and *Channa.*

The blood space is bound by the flanges of the pillar cells throughout the surface of the lamellae except outer margin. There are mucous glands with typical goblet cells that provide a thin film of mucous to prevent dehydration of the respiratory epithelium. Basophilic mast cells are the reservoir of heparin, the anti-coagulant. Other cells found in secondary gill lamellae are acidophilic granular cells, chloride cells and bi- and tri-nucleate glandular cells. Columns of basement membrane material run transversely across the secondary lamellae connecting the membrane of the opposite sides. These columns are deeply enfolded by the pillar cells. The columns contain many collagen fibrils and together with the basement membrane provide structural support to the secondary lamellae.

The number and dimensions of the secondary lamellae vary between different fish species according to their activity. A fish of 1 kg weight have about 18,000 cm^2 of secondary lamellae and an active fish like tuna have the same up to 5 million. The presence of large number of secondary lamellae in fish is due to the fact that the O_2 content of water is low, and secondly, the O_2 diffusion rate is relatively low in animal tissues even though the secondary lamellae are thin-walled to minimize diffusion distance. In active fish, like mackerels, the total lamellae surface exposed to water may be as high as 1000 sq.mm/gm body weight. The epithelial layers, basal lamina, collagen and the endothelial/pillar cell flanges together form the physiological barriers between blood flowing inside the lamellae and the outside water bathing it. The total thickness of the blood-water barrier may be 1 to 5μm.

Fishes like *Wallego attu, Catla catla, Hilsa ilisa* have one each of afferent and efferent branchial vassels in each gill arch but in others (*Labeo rohita, Clarias batrachus, Anabas testudineus*) there are two efferent branchial vessels in each arch (Dutta Munshi, 1960). Blood flows in the lamellae opposite to that of the water current ventilating them (counter-current) for maximum exchange of gases (up to 80%). This counter-current device is also useful for heat and salt exchange. Each afferent vessel runs through the entire length of the gill arch and produces number of branches and capillaries to the primary and secondary gill lamellae, gill rays. Gas exchange takes place in the capillaries. These capillaries finally join to form a short vessel, which carries the blood to the primary efferent vessel running along the margin of the primary gill lamellae. This vessel collects blood from the secondary lamellae of both sides and carries the oxygenated blood to the main efferent branchial vessel of the gill arch. Blood in the secondary lamellae flows through a sinus, the lining of which is made by the flanges of the pillar cells.

5.1.1.2. Gas exchange

The functional gill area for gas exchange can be reduced by re-routing the blood flow to non-lamellar pathways; thereby the respiratory gill area becomes non-respiratory. This happens in aged fishes when O_2 demand is less. In eels and dogfish (*Scyliorhinus*) the efferent and afferent arteries are connected in gill filaments and there are other links between the afferent arteries and the central venous space of the gill filament. Through these connections some blood flows to non-respiratory route.

The efficiency of the gill depends on the O_2 diffusion capacity. The diffusion capacity is the ratio of volume of O_2 transported across the blood-water barrier at the gill surface in unit time to the mean difference between the partial pressure of the gas in the water and blood. A riverine active or resting fish requires about 5 ml and 15 to 20 ml O_2 per hour per 100 gm body weight, respectively. For this amount of O_2, it has to pump 15 to 30 ml. of water to ventilate the gill surface per minute. According to Fick's law of diffusion, the rate of diffusion of O_2 across the gill is determined by the gill area, the diffusion distance across the gill epithelium, the diffusion constant and the difference in PO_2 across the gills (Crompton *et al.*, 2003). So, O_2 concentration is the most appropriate term for expressing O_2 levels in water. But O_2 concentration is the more commonly used term and for a given temperature and salinity, the PO_2 and O_2 content of water is linearly related.

5.1.1.3. Gill ventilation

Gill ventilation is the amount of water that the fish is able to pump across the gill lamellae for gas exchange. It can be measured as the amount of water per minute per gram body weight. Sometimes an amount of water flows out without any gas exchange. This amount of water is called dead space water. These dead spaces act as water shunts, allowing direct passage of water across the gills without O_2 diffusion. Axial flow of water (away from gill surface) contributes to the physiological dead space more than the inter-lamellar flow (nearer the gill surface). Dead space increases with greater velocity of flow.

There is an unidirectional flow of water over the epithelial surface of the gill for gas exchange (O_2 in and CO_2 out), where there is a counter-current mechanism between blood and water. The reason for this type of water flow is the energetic nature of the system. The energy that would be required to move water into and out of a respiratory organ would be very high than that used to move air because water holds low O_2 (Groot *et al.*, 1995). Most of the O_2 from water is taken in by the blood because the diffusion gradient is kept high by the blood. There are two ways for the ventilation of gill in fishes- active ventilation by the use of two pumps and passive or ram ventilation.

A. **Active ventilation :** The elasmobranches and teleosts use two pumps to ventilate their gills. The buccal or opercular pump is the pressure pump, which works upstream to the gill resistance and a suction pump that works downstream. The respiratory cycle in fish can be divided into three consecutive phases using the two pumps. In phase-1, the coraco-hyoid and coraco-branchial muscles contract to widen the angle enclosed by the gill arches. Thus the orobranchial (above the mouth) cavity is reduced and pressure within it increase. Water enters by suction through the mouth cavity and/or spiracle and during this phase the gill flaps are held to the skin by external water pressure. So, the external slits are closed. After an initial increase (as water flows out of the orobranchial cavity), the parabranchial cavity (below the mouth) also decreases in volume and pressure within it also increase. In phase-2, abductor muscles of the lower jaw and gill arches relax, pressure within the orobranchial cavity decreases as it expands passively. The adductor muscles (inter-arcular adductor) between the upper and lower portions of each gill arch contract and the mouth cavity functions as a pressure pump. As the oral valve prevents forward flow of water out of the mouth, the water is directed backward towards the internal gill clefts. Pressure at the inner gill surfaces is reduced and water is drawn into the gill cavities, that are still closed. In phase-3, the adductors relax, the internal gill clefts become narrower and the water is forced through the gill lamellae. Then the external gill clefts open passively and allow the water to flow outside. Before the next cycle begins, there is a rapid expansion of the orobranchial cavity at the end of the phase-2 during which the hypobranchial muscles are active.

B. **Passive (Ram) Ventilation :** In fishes like sharks, which swim continuously, the branchial pumps are not in use above a certain swimming speed as the forward motion of the fish provides sufficient ram (passive) gill ventilation. These fast swimming fishes ventilate their gills only keeping their mouths open and allowing the water to flow into the gills. The change takes place when the pressure difference across the gills reaches 2kPa. Ram ventilation is less energetically costly compared to the active ventilation. Ram ventilation requires nearly 9% of the total energy budget but branchial pumping requires about 15% of the total energy. The steady exist of water from the opercula during ventilation helps to maintain the boundary layer, producing a better flow regime with less drag.

5.1.2. Nerve Supply to Gills

Fish use cranial muscles for gill ventilation. These muscles are innervated by a dorsal group of cranial nerves existing from brain stem and termed as branchial nerves. This series of nerves contain sensory fibers and in most cases visceral motor components. The nerves innervating respiratory muscles include trigeminal V^{th}, which provides the major innervation to the mouth of all vertebrates, including the maxillary branch to the upper jaw and madibular branch to the lower jaw, responsible for motor control of the jaw-closing muscles. Jaw opening is passive in gill ventilation. The facial VIIth nerve provides the hyomandibular branch to the branchial muscles in the hyoid arch and in teleosts, the opercular muscles. The glossopharyngeal IXth and the vagal X^{th} cranial nerves innervate the gill arches and provide afferent innervations of the mechanoreceptors and chemoreceptors important in ventilating control and efferent innervation to intrinsic respiratory muscles in the gill arch. These branchial nerves have their efferent cell bodies and afferent sensory projections located dorsomedially in the brain. A longitudinal strip of neurons with spontaneous respiration-related brusting activity extends dorsomedially throughout the whole extent of the medulla (Hoar and Randall, 1970). These neurons make up the elements of the trigeminal V^{th}, facial VIIth, glossopharyngeal IXth and vagal X^{th} motor nuclei that drive the respiratory muscles together with the descending trigeminal nucleus and the recticular formation. The intermediate facial nucleus, receives vagal afferents from the gill arches that innervate many tonically and physically active mechanoreceptors and chemoreceptors. The areas in midbrain such as mesencephalic tegmentum have efferent and afferent connections with the reticular formation. The respiratory rhythm apparently originates in a diffuse respiratory pattern generator in the reticular formation.

5.1.3. Accessory Respiratory Organs

Many fish can breathe air via a variety of mechanisms. The skin of Anguillid eels may absorb O_2 (Kramer and Mc Clure, 1982). The buccal cavity of the electrical eel may breathe air. Catfish of the families Loricariidae, Callichthyidae and Scoloplacidae absorb air through their digestive tracts. Lungfish, with the exception of the Australian lungfish, and bichirs have paired lungs similar to those of tetrapods and moist surface to gulp fresh air through the mouth and pass spent air out through the gills. Gar and bowfin have a vascularized swim bladder that functions in the same way. *Loaches trahiras* and many catfishes breathe by passing air through the gut. Mudskippers breathe by absorbing O_2 across the skin. A number of fish have evolved so called accessory breathing organs that extract O_2 from the air.

Labyrinth fish, such as gouramis and bettas, have a labyrinth organ above the gills that performs this function. Some other fish have structures resembling labyrinth organs in form and function, most notably snakeheads, pikeheads and the clarridae catfish family.

A. **Modified gills -** *Clarius batrachus* has modified branched gills with thick, widely spaced lamellae on the dorsal side of the filaments. Dendritic structures arise from the second and fourth gill arches, and remain in parabranchial cavities (Jordan, 1976). The thickened and bulbus shaped modified gills ensure adequate support in air. The gills remain properly moist.

B. **Skin -** The skin is highly vascular and serves for exchange of gases in *Anguilla, Amphipnous* and *Perioptlhalmus* when they are out of water during migration through damp vegetation (Berg and Steen, 1965). Desiccation of the body surface is avoided by limiting terrestrial sojourns to nocturnal movements through moist grass. The glandular secretions of the skin protect it from desiccation.

C. **Mouth -** The buccopharyngeal epithelium in electric eel (*Electrophorus electricus*) in contrast to true eels is supplied by large number of capillaries to make it highly vascular for acquiring required O_2. Whereas this region has a large surface area from surface convolutions and papillae. The tongue projects into the buccal cavity and pharynx to make it an efficient respiratory organ. The gills have degenerated.

D. **Gut -** One or more sac-like diverticulae develops from the pharynx of fishes, particularly of Channidae family and cuchia eel. These diverticula store air for a certain period for the purpose of gas exchange. The diverticulae have folded respiratory epithelium with a rich blood supply. Pharyngeal pouches or supra-branchial cavities appear in the roof of the buccopharynx. These cavities are lined with vascular respiratory epithelium with folds showing respiratory islets. The cavities open into the pharynx. These cavities posses some alveoli also. The pharyngeal diverticula play important role in respiration. The gill lamellae are much reduced. The epibranchial of the first gill arch is flattened and covered by a thin vascular respiratory membrane. It also have folded and branched respiratory labyrinthine as accessory respiratory system.

In some fishes either the stomach or the intestine is specially modified to help in aerial respiration. The inhaled air is swallowed and forced back into the alimentary canal and is stored for sometime in a special part of it. After respiratory exchange, the used up air is either passed out to the exterior through anus or is expelled through the mouth. The stomach or intestine have thin wall. The inner surface lined by a single layer of epithelium and devoid of mucous glands.

E. **Modification of opercular chamber -** In some species, the opercular chamber becomes bulged out like balloons to store air for sometime and then the air comes out through small branchial opening. The chamber has thin vascular wall for gas exchange. In more specialized air-breathing fishes sac-like diverticulae develop from the dorsal surface of the operular chamber. In *Heteropneustes fossilis* paired, thin-walled tubular sac-like structures extend backwards from the gill up to the middle of the caudal portion. These sacs are formed by the modifications of the primary and secondary gill lamellae. In *Anabas testudineus* the air chambers are specious and communicate with buccopharyngeal cavity.

F. **Air bladder -** The air (gas) bladder is also known as swim bladder. Even though it is the hydrostatic organ in fish to maintain body buoyancy, it has several other functions such as accessory breathing organ, sound production, and may also serve as a fat storage organ in some deepsea fishes. It remains between the vertebral column and digestive tract. It develops as a diverticulum of the anterior region of the alimentary canal. It may be connected to the esophagus by a pneumatic duct (physostomous type) or the duct may be absent (physoclistous type) but the bladder becomes a closed sac. All teleosts are physostomous at young stage and then become physoclistous. The bladder acts as a supplementary respiratory organ. The air-breathers may be facultative or obligatory. The former have highly sacculated or alveolar lining in the gas bladder. These fishes can servive in water devoid of O_2 if enabled to swallow air which then pass into the gas bladder. Since the bladder in physostomous fishes always contains more CO_2 than the air, it has been suggested that elimination of this waste gas is performed there also.

The bladder is supplied with blood from dorsal aorta (in *Lepisosteus*) or from branchial vessels (in *Amia*). But in most fishes blood comes from the coeliaco-mesenteric artery. Blood returns to heart from the gas bladder through post-cardinal veins. The air-bladder is modified for aerial respiration acting as a lung in Dipnoi and others. The network of blood capillaries covered by a single layer of epithelium facilitates diffusion of gases.

5.1.4. Gas Transport

The effectiveness by which O_2 is taken up from water depads on the efficiency of the animal's respiratory pigment, such as hemoglobin (Hb). At high O_2 concentrations Hb combines with O_2 to from oxyhemoglobin (HbO_2). The amount of O_2 that combines with blood at equilibrium at a given PO_2 can be plotted as an O_2 combining curve. Several factors affect the affinity of Hb for O_2, but two of the most important factors are CO_2 and temperature . Their influence is complicated by diverse physiological requirements and tolerance levels among spices and also by size and age differences among animals or the same species.

Nitrite in the blood oxidizes Hb to methemoglobin, which is incapable of transporting O_2 (Jaffe, 1964). Methemoglobin in fishes can be detected by the color of the blood and gill, which turn brown.

In blood CO_2 is transported as bicarbonate (HCO_3^-). The HCO_3^- moves from the blood by passing through the RBC in which O_2 also binds to Hb at the respiratory surface, causing hydrogen ions (H^+) to be released. The increase in H^+ ions combines with HCO_3^- to from CO_2 and OH^-. So, more CO_2 is formed and can leave the blood across the respiratory surface. Excess H^+ ion binds to OH^-, forming water and allowing the pH to increase to facilitate the binding of O_2 to Hb. The release of O_2 from Hb in the tissues makes the Hb available to bind to H^+, promoting the conversion of CO_2 to HCO_3^-, which helps to draw CO_2 from the tissues. Thus, CO_2 that is being transported into and out of the RBC minimizes changes in pH in other parts of the body because of proton binding and proton release from Hb, as it is deoxygenated and oxygenated, respectively. The O_2 level in water should be kept at or a little higher for fish culture. The recommended minimum dissolve O_2 requirements are as follows :

1. Coldwater fish - 6 mg per liter (70% saturation).
2. Tropical freshwater fish - 5 mg per liter (80% saturation).
3. Tropical marine fish - 5 mg per liter (75% saturation).

The amount of free CO_2 in solution is a function of pH. As pH decreases, the CO_2 concentration increases. The sources of CO_2 in water are the dissociation of bicarbonate ions. The presence of free CO_2 in blood influences the shape and position of the O_2 combining curve. Generally acidity increases as the cells metabolize and release CO_2 into the blood. As the blood CO_2 level increases, the affinity of a respiratory pigment for O_2 is reduced and O_2 is released to the tissues more readily than it would be otherwise at comparable partial pressures of O_2. As a result, the O_2 combining curve is shifted to the right. This shift is called Bohr Effect. Ectotherms show varied responses to the Bohr Effect. Many invertebrates show no response at all. Lenfant and Johansen (1966) could not find evidence of a Bohr effect in Pacific dogfish (*Squalus acanthias*). There is a marked Bohr effect in many other fishes as blood CO_2 increase (Saffron and Gibson, 1976).

In the blood of some fishes, there is much decline in the amount of O_2 that the respiratory pigment can hold at high levels of CO_2. In such cases, the fish's O_2 combining curve is shifted so far to the right that it falls outside the useful range and the animal suffocates, even in water that is saturated with dissolved O_2. This phenomenon is called Root effect (Root, 1931). Hoar (1975) pointed out that the relationship between free CO_2 and the O_2 combining properties of blood vary with the animals ecological condition and the adjustments that it must make to deliver O_2 to the tissues under varying environmental conditions. Animals with high Bohr effect have respiratory pigment with a low affinity for O_2. Increased temperature weakens the bond between Hb and O_2.

5.2. RESPIRATION AND GAS EXCHANGE IN SHELLFISH

Some crustaceans have a high surface to volume ratio, and gas exchange across the integument is almost sufficient for respiration, particularly in branchiopods and other groups that have foliaceous trunk limbs. The carapace, when present also offers a substantial surface area and is an important site for gas exchange in small crustaceans and larvae. As body size increases, the ratio of surface area to metabolic tissue decline, while at the same time the general integument (exoskeleton) increase in thickness supporting the larger body mass. To maintain adequate levels of gas exchange under these conditions, the larger crustaceans have developed gills which increase the thin gas-permeable cuticle area without compromising general cuticular function.

5.2.1. Gill

Many small crustaceans, such as copepods, have no special respiratory organs. Gas exchange takes place through the entire thin integument. The inner wall of the carapace, towards the trunk, is often rich with blood vessels and may in many groups be the only respiratory organ. Gills, when present are formed by modifications of parts of appendages, mostly the epipodites. These thin-walled lamellate structures are present on some or all of the thoracic appendages in cephalocarids, fairy shrimps and many malacostracans. In mantis shrimps (order Stomatopoda), gills are found on the exopodites of the pleopods. In euphausiids, a single series of branched epipodial gills are fully exposed. The decapod gills, protected by the over-hanging carapace are arranged in three series at or near the limb bases.

5.2.1.1. Gill number

Primitively, there is four gill on each side of every thoracic segment. The gills arise from the body wall or near the point of attachment of the appendage and have been given different names depending upon their position. If all the gill series were present on all segments, there would be a total of 32 gills on each side of the body, but no decapod has retained this maximum number. The penaeid shrimp, *Benthesicymus*, with 24 gills has the greatest number, but reduction to far less than this is the general rule. The snapping shrimp, *Alpheus*, with six gills on each side, has one gill on the third segment (maxilliped) and one gill on each of the leg segments. The lobsters, *Homarus*, has 20 gills on each side of the body distributed as - one gill on second, three on third maxilliped and four on each of the legs except the first, which have three and the last leg bears only one gill. Nine pairs of gills are usually found in marine crabs. However, the swimming crab, *Callinectes*, has eight gills on each side. There are three gills on each side in little pea crab, *Pinnotheres* - one on the third maxilliped and two on the chelipeds. In all decapods, the gills of the first maxilliped are vestigial or even absent.

All decapods posses gills (branchiae), except dendrobranchiate shrimp, *Lucifer*. The gills are known as tricobranchs. The number and arrangement of gills varies depending on the species, but typically four gills are attached to some or all the thorocic somites. One gill blanket, the pleurobranch, is usually attached to the lateral wall of the somite dorsal to the articulation of the walking leg. Two gills, the arthrobranchs, are attached to the arthrodial membrane (the cuticle joining the limbs to the body) between the coxa and the body wall. The remaining one gill, the podobranch, projects from the coxa of the walking leg (periopod) and are attached to the epipodites of these limbs (Calman, 1909). In addition there may be a leaf-like lamella or mastigobranch, accompanying the podobranch and a tuft of coxal setae known as setobranchs. These are filamentous gills occurring on the basal joints. The arrangement of the gills on the thoracic somites, walking legs and mouth parts is termed as the branchial formula (Table-5.1), which is commonly used in most moden species description of decapods.

Table 5.1 : The distribution of the gills on the thoracic segment.

Thoracic segment	I	II	III	IV	V	VI	VII	VIII
Pleurobranchs	X	X	X	X	X	1	1	1
Arthrobranchs	X	1	2	2	2	2	2	X
Podobranchs	X	1	1	1	1	1	1	X
Epipodites	3	3	3	3	3	3	3	X
	1st	2nd Maxilliped	3rd	Great chelae	Walking legs			

3 → Epipodite present X → Both epipodites and gill absent

5.2.1.2. Gill design

The gills are not visible to the outside as they remain in a gill (branchial) chamber covered by the branchiostegite on each side of the cephalothorax. The branchiostegite is also referred as gill chamber. The inner linings of this are soft, membranous and highly vascular with minute blood lacunae. These form large respiratory surfaces for gas exchange. To ensure that oxygenated water reaches the gills, the branchial chambers are ventilated by specially modified bailers (scaphognathites) on the 2nd maxillae. Typically the gills are out growths from thoracic legs (epipodial gills), although certain species also utilize the pleopods and some others have developed specialized pleopodal gills. The axial stems of the podobranchs perhaps represent modified epipodites. The epipodites are three pairs of simple, foliaceous and highly vascular outgrowths of the integument, given out from the coxal segments of the three pairs of maxillipeds. They occupy the anterior part of the gill chamber (Fig-5.2). The epipodites of the first pair are bilobed and larger than the other two. The epipodites also serve as respiratory organs.

It has been suggested that the arthrobranches and pleurobranches are derived from the epipodites of precoxae which have fused with the body (Brown, 2002). Mastigobranches, rather than the gills, most closely resemble the epipodites.

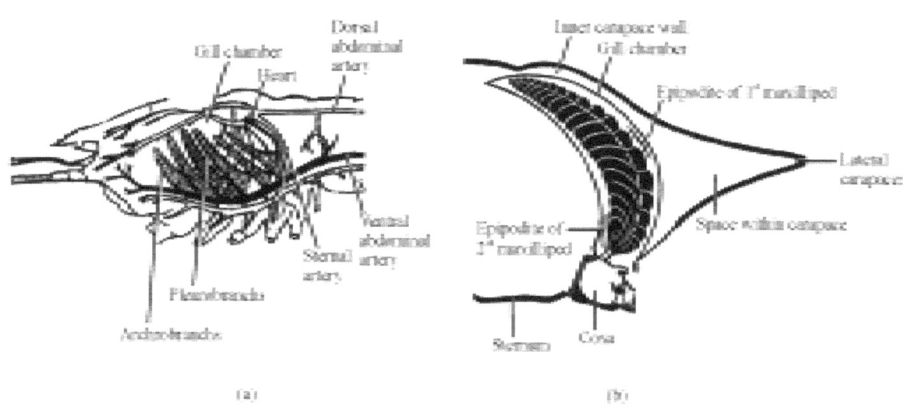

Fig. 5.2 : Gill of prawn (a) and crab (b).

Each gill consists of numerous filaments arranged around a central axis like a bottle brush. The gills of decapods are in a branchial (gill) chamber under the carapace, and oxygenated water enters through them. The epipodial gills of syncarids and euphausiids are unprotected as the carapace is either lacking or does not cover the leg bases. In amphipods, the gills are usually simple sacs or plates, and are oxygenated by, the pleopods. In stomatopods and isopods, gill-like outgrowths of the pleopods are the main organs of respiration.

In all decapods, three distinct gill types (Fig-5.3) such as dendrobranchiate, trichobranchiate and phylobranchiate, have evolved. Intermediate forms of the above types are not uncommon throughout the decapods (Calman, 1909, Felgenhauer and Abele, 1983). The dendrobranchiate gill, with branched filaments, is unique to dendrobranch prawns and sergestoid shrimps. These gills have paired complex lateral branches, arising from the central brachial axis; With a series of subdivided secondary rami or filaments from each lateral branch. The tricobranchiate gills with characteristic serial tubular filament arise from central brachial axis. This type of gill is found in crayfish and rock lobsters. The phyllobranchiate gills of crabs and eukyphid prawns exhibit flat paired lamellar branches extending from the branchial axis.

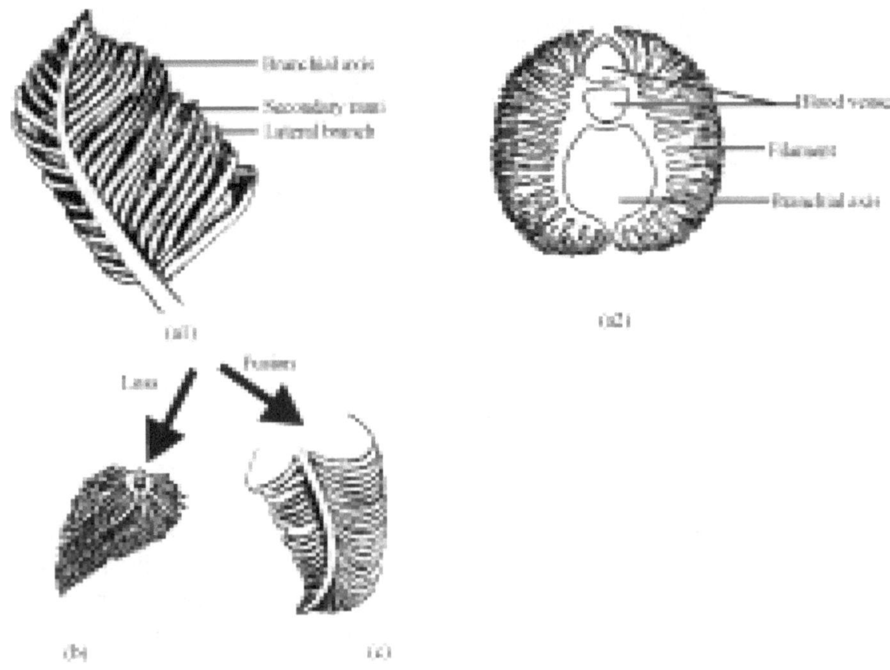

Fig. 5.3 : Types of decapod gill - dendrobranchiate (total-a1, transverse secion -a2) trichobranchiate (b) phyllobranchiate (c) gill.

Decapod branchial epithelia contains a minimum of six types of cells such as chief cells, pillar (or pilaster or trabecular) cells, striated cells, glycocytes, nephrocytes (or podocytes) and granular cells (Johnson, 1980; Goodman and Cavey, 1990). The supportive pillar cells provide structural facilities for efficient blood flow (Johnson, 1980: Ciofi, 1984). The striated cells are mostly found near the excurrent hemolymph channel for ion regulation (Goodman and Cavey, 1990). These three cells make contact at same point with the endocuticle of the gill lamella. Nephrocytes are phagocytic, exhibiting interdigitating foot processes that attach to branchial membrane. These cells filter hemolymph via pedical pore diaphragms and the basal lamina. Sequestered substances are enclosed within vacuoles in these cells (Foster and Howse, 1978, Johnson, 1980: Goodman and Cavey, 1990). Glycocytes and granulocytes, the complex fibrillar aggregates are packed within glycogen molecules (Foster and Howse, 1978: Goodman and Cavey, 1990). The functions of these six cells are little known, other than storage.

Gas exchange in crabs occurs across the surface of gills that hang into the branchial chamber on either side of the body. Eight pairs of gills provide gas exchange when immersed. Two pairs on each of maxilliped 2 and 3 provide about 30% of the total surface area, while the two pairs on walking leg 1 provide 43% and one pair on each walking legs 2 and 3 provide the remainder. Seawater is pumped into and through the chamber by the beating of special paired appendages, the gill bailers or scaphognathites. Seawater enters the branchial chamber at the back and over each of the walking legs. A fringe of bristles hangs down from the lower edge of the carapace for protection against entry of particulate material and unwanted organisms. The branchial chamber is enlarged with increased structural support for the gills as a terrestrial adaptation in land crabs for aerial respiration and gas exchange. Adaptations for gas exchange directly from air include reduction in the surface area, number and size of gills and a well vascularized lung area within the branchial chamber for keeping the gill moist. But in tropical and subtropical regions, where semiterrestrial life of crabs is more prevalent such adaptions are common.

5.2.2. Blood Supply to the Gills

Three longitudinal blood channels run through the gill base along the entire length of the gill. Two are lateral channels running along each lateral side. The third one is median longitudinal channels, running through the apex of the gill base, beneath the outer median groove of the gill. Lateral channels are connected together by a series of transverse connections. In each gill, the lateral channels of that side gives off a slender marginal channels, which runs all along its margin and lastly joins with the median channel. In the axis of each gill runs an afferent and an efferent branchial channel. Deoxygenated blood from body is brought to the gill by the afferent channel and then back into efferent channel. A longitudinal transverse connective in each filament divides the two channels. Blood flows from afferent channel into longitudinal channels. Passing through the marginal channels, it reaches the median longitudinal channel. During this flow blood is oxygenated. From median channel blood goes to efferent channel. The blood from the sternal sinus flows through the gills on its way back to the heart. In the brachyurans, blood flows through the lamellae within a fine sinus network. The cardiovascular shunting during walking activity of crabs leads to decreased flow of blood to the digestive system and increased flow to muscles of the walking legs and respiratory system. Ventilation rate also doubles during walking.

5.2.3. Gas Exchange

Gases diffuse across the respiratory surface. Since the chitinous material of the body wall is relatively impermeable, special mechanisms have evolved to boost O_2 intake. These include increased surface area, highly vascularized respiratory surfaces, ventilating mechanisms (current-directing exopods and bailer plates of maxillae and maxillipeds) and presence of blood pigment hemocyanin. The branchial cuticle may vary greatly in its thickness, depending on whether gills are anterior or posterior in the branchial chamber. Based on morphology the anterior gills may have respiratory function and the posterior lamellae serve as ion regulator (Towle abd Kays, 1986; Goodman and Cavey, 1990) because of thicker epithelial areas in the later. These areas have extensive infoldings of the basal-lateral membranes with many mitochondria. Other regions of the gill like branchiostegites of decapods have both the above functions.

The ventilating current is produced by the beating of the gill bailer. Water is pulled forward and the exhalent current flows out anteriorly in front of the head. In shrimps, the ventral margins of the carapace fit loosely against the sides of the body and water can enter the branchial chamber at any point along the posterior and ventral margins of the carapace. The macruran carapace is fitted more tightly, thus the entry of water is limited to posterior carapace edges and at the bases of the legs. In brachyurans, the forward position of the inhalent openings results in water taking a U-shaped movement through the gills. The bottom dwellers and burrowing species develop many mechanisms to prevent clogging of the gills. The bases of the chelipeds in crabs and the leg coxa of crayfish and lobsters bear setae that filter the incoming water. In crabs, the gills are cleaned by the fringed epipodites or by the gill bailer's intermittent reverse beating pattern. This reverse current is also used to ventilate the gills when the ventral body parts are covered with mud. But the forward current is generally used for gill ventilation. As a ventilating adaptation, the burrowing decapods have developed inhalant siphons. In some crabs and shrimps, the first antennae are held together forming an inhalant passage between them. In *Metapenaeus*, this is done by first antennae and the antennal scales and with the second antennae in burrowing crab *Corystas*.

When the animal is walking, the movements of the limbs produce backward and forward motions of the gills attached to the limbs and also of the epipodite. These movements are often sufficient to renew the water around the gills but when the animal is at rest, the 2nd

maxillae flutter rapidly. This causes water to come out of the gill chamber in a forward direction while freshwater enters between the legs. The extremely delicate and thin gill plates act as excellent permeable membranes for the exchange of gases.

Oxygen is not very soluble in saline fluid such as blood and the amount that can be carried in simple solution is restricted. Hemocyanine of decapod blood plasma increases the O_2 carrying capacity by a factor of 10 or more. It is a copper containing respiratory pigment of molecular weight 9 x 10^6d. It is colorless when deoxygenerated and pale blue when oxygenated. Hemocyanine binds with O_2 at high O_2 tensions present at the respiratory surface and releases it at the tissues with lower tensions. Branchiopods, ostracods and copepods use Hb rather than hemocyanine and can increase its concentration in the blood in low O_2 environments.

CHAPTER - 6

Excretion and Osmoregulation

Excretion is the process of elimination of metabolic wastes that include toxic nitrogenous end products of the protein metabolism and the gas such as carbon dioxide (CO_2). These nitrogenous wastes are excreted through urine. Fecal matter is not included in the excretory products as it is simply the undigested food materials, and are not metabolic end products. The membranes of aquatic animals, that is permeable to gases, pass water and solutes between environment and body. Most aquatic animals maintain in their bodies a reasonably constant proportion of water and solutes. When it is different from the surrounding water, it requires some mechanism for regulation, which is known as osmoregulation. Most invertebrates in the sea have nearly the same salt concentration (isosmotic or isotonic) as the environment. Sharks and rays (elasmobranches) tend to have a slightly higher concentration of salts and urea (excretory end products) than seawater but the body fluids of most teleosts (bony fishes) have much lower salt concentration than seawater. Freshwater fishes, which live in a medium of lower salt concentration (hyposmotic or hypotonic), have the threat of hydration, so they have special mechanisms to get rid of excess body water. Marine fishes that live in an environment of high salt concentration (hyperosmotic or hypertonic) have the regulatory mechanisms by which excessive loss of water from the body is prevented and body osmolarity (proper osmotic concentration) is maintained by retaining less toxic nitrogenous excretory products. Marine fishes, which lose water osmotically across the gills and must get rid of salt, tend to drink large volume of seawater and produce very small quantities of urine (Fig-6.1). The different salts from the water do not remain in the fish body in proportion to their abundance in the environment. Most animals are able to concentrate some ions while excreting the others. Marine animals usually concentrate sodium, potassium and chloride ions. Thus, the physiology of excretion and osmoregulation are closely associated and urination is a mechanism of regulation for both the processes.

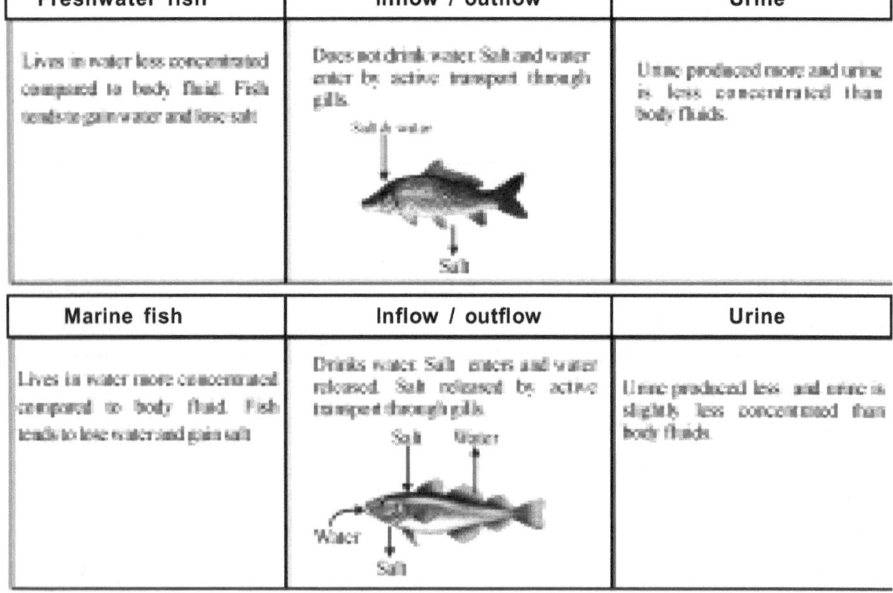

Freshwater fish	Inflow / outflow	Urine
Lives in water less concentrated compared to body fluid. Fish tends to gain water and lose salt	Does not drink water. Salt and water enter by active transport through gills. Salt & water / Salt	Urine produced more and urine is less concentrated than body fluids.
Marine fish	**Inflow / outflow**	**Urine**
Lives in water more concentrated compared to body fluid. Fish tends to lose water and gain salt	Drinks water. Salt enters and water released. Salt released by active transport through gills. Salt / Water / Water / Salt	Urine produced less and urine is slightly less concentrated than body fluids.

Fig. 6.1 : Inflow and outflow of salt and water of different fishes in different osmotic environment.

6.1. EXCRETION AND OSMOREGULATION IN FISH

6.1.1. Excretory Waste Products

Carbon dioxide and metabolic water produced in animal during metabolism can easily diffuse into the environment from respiratory surfaces. Nitrogenous waste excretion is more difficult, yet necessary. When proteins are catabolised, the amino group is removed. In the body, the amino group is quickly oxidized to form ammonia (or at high body pH the ammonium ion). Ammonia (NH_3^+) is highly toxic and highly soluble in water. Elevated NH_3^+ level in the body can lead to health problem, and even death. Ammonia can also adversely affect both metabolism and amino acid transport. Excessive amounts of NH_3^+ in the system increases body pH, which causes changes in the tertiary structure of proteins, and thus their cellular functions.

Most fish release their nitrogenous waste as NH_3^+. So, they are known a ammoniotelic. The aquatic organisms, particularly in freshwater, face no problem to excrete NH_3^+, which can diffuse passively out of the respiratory structure such as gills in a concentration gradient between the organism (higher level) and the environment (lower level). During water deprivation or aestivation, the NH_3^+ is converted to less toxic but more complex urea and uric acid (Fig- 6.2).

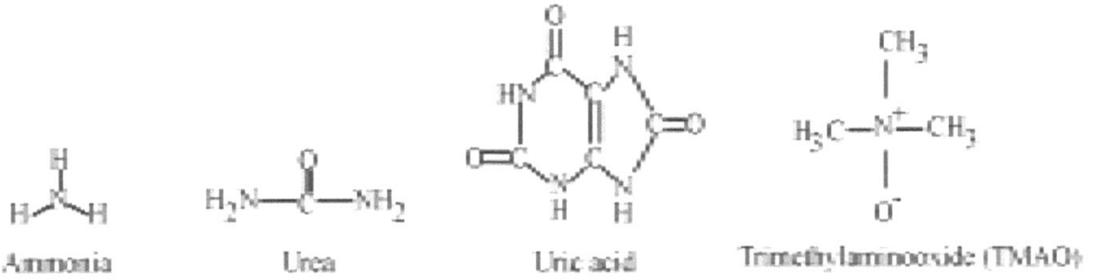

Ammonia Urea Uric acid Trimethylaminooxide (TMAO)

Fig. 6.2 : Different excretory waste products.

Ammonia carries off one atom of nitrogen, urea two and uric acid four. The synthesis of urea and uric acid requires energy, which is not required for NH_3^+ production. Teleosts produce urea from NH_3^+ via ornithine-urea cycle in liver and via other pathways (Fig- 6.3). The animals who produce urea and uric acid are known as ureotelic and uricotelic, respectively. Uric acid excretion in fishes is very less because it is largely insoluble in water and precipitates. Urea being less toxic can accumulate in the blood to some extent as in sharks for overall osmotic balance. Sharks use it in their blood to make them hyperosmotic in relation to seawater, thus they tend to gain water from the ocean and do not have to worry for dehydration.

Other nitrogenous end products are trimethylamine oxide (TMAO), amino-*N*, creatine, creatinine and inorganic electrolytes, of which TMAO (Fig-6.2 and 6.3) is accumulated in blood in marine species for osmotic balance. Some of the wastes diffuse through the gills. Blood wastes are filtered by the kidneys. Salt water fishes tend to lose water because of osmosis. Their kidneys return water to the body. The reverse happens in freshwater where the kidneys produce dilute urine. Some fish have specially adapted kidneys that vary in function, allowing them to move from freshwater to salt water and *vice versa*.

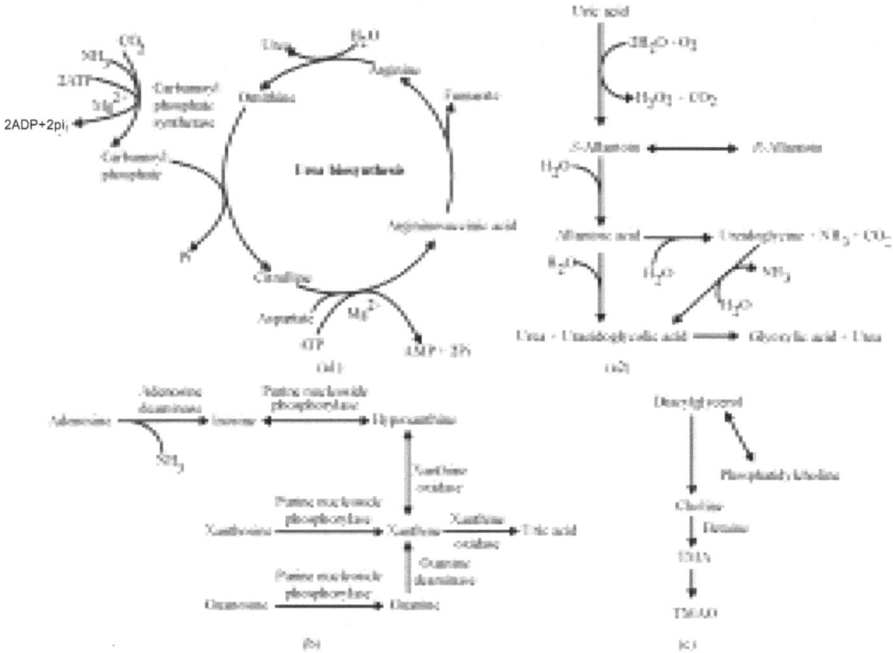

Fig. 6.3 : Bio-synthesis of urea (a1-a2), uric acid (b) and TMAO (c).

6.1.2. Major Structures Involved in Excretion and Osmoregulation

Organisms in different environments use different body parts for the process of excretion and osmoregulation. The major organs involved are the integument, respiratory surfaces, kidneys, salt or rectal gland and chloride cells. All animals use at least one of these structures for the said processes. The common characteristic in organs like gills, skin and kidneys is the presence of anatomical and functional polarized cells, with transport epithelia, which determine the osmoregulatory capabilities of the structure, because of their permeability to various solutes and water.

The integument functions in osmoregulation by acting as a barrier between the extracellular compartments and the environment to regulate water gain and loss, as well as solute flux. The permeability of the integument to water and solutes varies from species to species, and animal to animal. Respiratory surfaces such as the alveoli of the lung (in terrestrial species) and gills (in aquatic organisms) also have role in excretion and osmoregulation. Respiratory surfaces are the major sites for the elimination of CO_2, other gaseous wastes and water.

The kidneys are the main organs, in vertebrates including fish, involved in maintaining osmotic balance and excreting harmful metabolic substances.

6.1.2.1. Kidney structure

The kidneys (Fig-6.4) are paired, elongated structures, placed above the alimentary canal and are close to vertebral column. The hagfishes, which are not vertebrates but are among the most primitive living chordate, have kidneys with segmentally arranged excretory tubules. This suggest that the excretory segments of vertebrate ancestors were segmented. However, the kidneys of most vertebrates are compact, non-segmented organs containing many uriniferos tubules, arranged in a highly organized manner. The vertebrate excretory system includes a dense network of capillaries closely associated with the tubules along the duct and other structures that carry urine out of the kidney tubules.

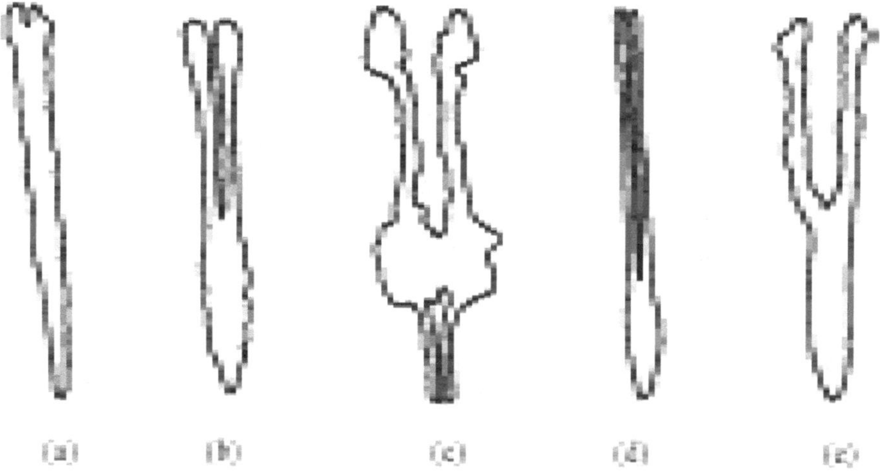

Fig. 6.4 : Kidney of different fishes, *Salmon* (a), *Plecoglossus* (b), *Cyprinus* (c) *Anguilla* (d) and *Seriola* (e)

The kidney has two parts - head and truck kidney. Fish kidney is made up of many individual functional units known as nephrons or uriniferos tubules., each containing of a renal corpuscle or Malpigian body (as per the name of the Italian physician and biologist Marcello Malpighi, 1628-1694) and renal duct or tubule with various terminal modifications (Fig-6.5).

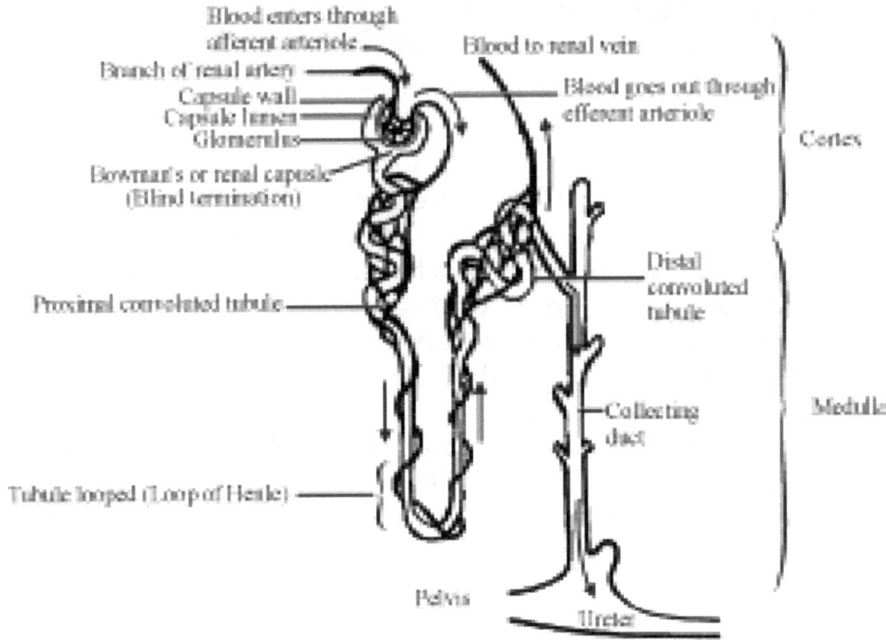

Fig. 6.5 : Diagram of renal corpuscle.

The Malpigian bodies are lacking in some fishes, and such type of kidney is known as aglomerular. The tubules join to a collecting duct called archnephric duct or ureter located posterior to cloaca. The kidney is holonephros when it extends to the entire length of the body. Such kidney is found in larvae of some cyclostomes. In fish the most anterior tubules have been lost, some middle tubules are associated with testes and there is a multiplication of tubules posteriorly. Such kidney is known as ophisthonephros. Generally in fishes the tubules of the anterior region become functional in early life and such type of kidney is designated as pronephros. Fish kidney has pronephric head region (non-functional in terms of excretion) and mesonephric (middle region functional) trunk region. The mesonephros serves as the main excretory organ for aquatic vertebrates like fish. The anterior tubules of mesonephros are usually reduced and converted to lymphoid organ, and the posterior tubules take up excretory function throughout the life. The mesonephros is called as the Wolffian body after Caspar Friedrich Wolff, who described it in 1759. The mesonephros is composed of the mesonephric duct (also called Wolffian duct), mesonephric tubules and associated capillary tufts. A single tubule and its associated capillary tuft is called a mesonephric excretory unit. These units are similar in structure and function to nephron of higher vertebrates. The mesonephros has mesodermal origin in vertebrate embryo. In higher vertebrates (mammals, including man), the kidneys is metanephric or posterior kidney.

Pronephros function is generally lacking in major adult teleost (Romer and Parson, 1977). According to Ogawa (1961) head kidney may remain merged posteriorly with the ophisthonephros in herrings. In others the head kidney is distinguished and separated from ophisthonephros (*Cirrhinus, Anguilla, Notepterus, Clarias* etc.)

The head kidney is generally made up of lymphoid, hematopoietic tissues and hardly contains a few nephrons and collecting ducts. So, it has no excretory function. The inter-renal and chromaffin (suprarenal) tissues forming the endocrine gland, adrenal, are found diffused in the head kidney near the cardinal vein. The corpuscles of Stannius, another endocrine gland, is usually located on the dorsal side of the middle to posterior part of the kidney, which arises from embryonic pronephric duct. The trunk kidney contains a large number of nephrons. The number of nephrons in freshwater teleosts may be about 10,000 per kidney. In the marine fishes, however, there is a great variation in number, shape and size of the renal corpuscles. Large and well vascularized renal corpuscles are rarely found. In the renal parenchyma there are many non-myelinated nerves with large cell bodies (Bulger and Trump, 1968).

The nephron begins as a double walled blind cup called Bowman's capsule lined by visceral (inner) and parietal (outer) layers of epithelial cells (Fig. 6.5). Bowman's capsule is named after Sir William Bowman (1816-1892), a British surgeon. The inner layer lies just below the thickened glomerular basement membrane and is made of podocytes which send foot processes over the length of the glomerulus. These foot processes cross each other forming filtration slits. The size of the filtration slits restricts the passage of large molecules (eg. albumin) and cells (eg. RBC and platelets). The foot processes have a negatively-charged coat (glycocalyx) which limits the filtration of negatively-charged molecules, such as albumin. The Bowman's capsule encapsulates the well vascularized glomerulus with an inconspicuous mesangium. Attached to each Bowman's capsule is a long, thin tubule (which functions as dialysis unit) with three distinct regions - neck region, proximal (or first) and distal (or second) convoluted tubule or segment. The ciliated neck region is of variable length. The proximal tubule is lined by a single layer of cubical cells and have brush border, mitochondria and lysosomes. Its main role is to concentrate the salt in the interstitium, the tissue surrounding the loop. The proximal tubule links with distal tubule by the U-shaped loop of Henle. This loop have descending and ascending limbs, which play different roles in urine formation. The cells lining the distal proximal tubule contains many mitochondria which enables active transport using the energy of ATP. The distal tubules from many nephrons connect to a common collecting duct.

In minnow (*Dania*) prominent mucus secretion in the cell apices of the second part of the proximal tubule is observed. The kidney's second proximal tubule of sexually mature male stickleback (*Gasterosteus aculeatus*) secretes a mucus, which is used to glue together algal filaments on the formation of nest (Cleveland *et al.*, 1969). Beyond second proximal segment, there is a narrow ciliated intermediate segment which may be absent in some species. The distal segment, with clear cells and elongated mitochondria, joins with the collecting duct.

Glomerulus is a blood vessel tightly coiled with efferent and afferent arterioles and encapsulated by thin kidney cells forming Bowman's capsule. The blood supply to the glomerulus is from the renal artery arising from dorsal aorta whereas to the tubules from the renal portal system. The renal artery branches repeatedly forming arterioles to reach each nephron. The capillary bed of the tubules is also joined by the arteriole leaving the glomerulus. An afferent arteriole delivers blood to the glomerulus capillaries for filtration and an efferent arteriole drains filtered blood away from the same glomerulus. The efferent arteriole connects to a network of peritubular capillaries that are closely associated with the nephron tubule facililating reabsorption of water, ion and nutrients from the filtrate in the nephron tubule into it. Finally a renal vein drains the blood supply into post cava. The walls of the afferent arteriole contains the renin-secreting juxtaglomerular cells as an apparatus. These cells are the site of the enzyme renin synthesis and secretion and play a critical role in renin-angiotensin system. There is an close association between the blood vessels and the nephrons of the kidney. This association permits both extensive filtration from blood, and selective reabsorption back into blood.

Some teleosts have kidneys with frequent small glomeruli or a few species with medium sized glomeruli, others show a tendency for the glomeruli to become smaller. There are a number of marine fishes, including Antarctic bony fish, which are lacking glomeruli (eg. toadfish, *Opsanus*). Presence of antifreeze glycoproteins and aglomerularism of the kidneys probably play important roles in permitting Antarctic fish to adapt to near freezing water temperature. Since these small molecular weight glycoproteins are not filtered into the formative urine, the fish avoid the energy expenditure that may be necessary for their tubular reabsorption. Control of filtration and reabsorption takes place through hormonal action.

6.1.2.2. Kidney of different fishes

Adult hagfish have paired kidneys of which anterior and posterior parts are pronephros and mesonephros, respectively, with a degenerated middle portion in between the two (Cleveland *et al.*, 1969). Atubular glomeruli and aglomerular tubules are present in the middle portion. The pronephric kidneys contain single large Malpighian corpuscle. The Wolffian duct, which in young animals connects the pronephros and mesonephros, degenerates and the Bowman's capsule does not empty directly into tubules. The tubules has no direct connection to excretory canal. The mesonephros contains 30-35 large Malpighian corpuscles arranged segmentally on the medial side of an archinephric duct. Bowman's capsule is connected to the archinephric duct by a short neck. The dorsal aorta supplies segmental arteries to glomeruli. Capillary network of the ureters is served by post-glomerular circulation.

In Lamprey each triangular-shaped kidney has distinct glomerulus and tubules. Nephrons consists of renal corpuscles ciliated neck, a proximal segment and a convoluted segment without brush border (Policard, 1902). Efferent arterioles drain a sinus at the hilus, which is supplied by branches of the aorta. There is no renal portal system. The glomerulus (otherwise called glomus) contains capillaries that are widely open like freshwater fish. The straight and short neck segment is lined by columnar ciliated cells on two sides and on the other two sides by non-ciliated cuboidal cells. The first proximal segment, with many diverticula, contains many mitochondria and is cytologically similar to archinephric duct of *Myxine glutinosa* and the proximal tubule of mammalian kidney. The cells contain secondary lysosomes for intake and storage of colloidal materials (Schneidar, 1903), Convoluted segment is lined with cuboidal epithelium possessing mitochondria.

In elasmobranchs, nephrons are long with large glomeruli of widely distributed capillaries and thin mesangial areas (Kempton, 1956). Granulated arteriolar or juxtaglomerular cells are mostly absent but some open nephrostomes present (Bohle and Walvig, 1964). The nephron consists of neck, first and second proximal and distal segments and the collecting duct. The neck segment is long with a small thin-walled tubule covered by cuboidal epithelium. The first proximal segment is lined with tall cuboidal epithelium with brush border. It contains many mitochondria at the anterior region. The second region has many lysosomes. The second proximal segment have most nephrons. The distal tubule is lined with cuboidal cells without brush border. The collecting ducts are lined by columnar cells.

The kidney of bony fishes (teleosts) is with highest development in basic structural and functional patterns. Structurally advanced filtration-reabsorption device with at least six cytologically distinct tubular regions have developed in addition to the renal corpuscles as freshwater adaptation in advanced Teleostomi for the mono-valent ion-free urine production.

In marine teleosts, the filtration reabsorption system is modified into a tubular divalent ion secretary device. Ogawa (1961) had divided marine teleost's kidney into five types such as -

Type I - Two kidneys are completely joined along their entire length without any clear distinction between head and trunk kidney. Example - Herrings (Clupeidae).

Type II - The anterior portions of the kidneys are free but the middle and posterior portions are fused. The head and trunk kidneys are clearly distinguishable. Example - Marine catfishes (Plotosidae) and eel (Anguillidae).

Type III - The anterior and middle portions of the kidneys are free but the posterior parts are fused. The anterior portion has two thin branches. The head and trunk kidneys are distinctly separate. Most marine teleosts have this type of kidney. Example- Billfish (Belonidae), mullets (Mugilidae), mackerels (Scombridae), flounder (Pleuronectidae) etc.

Type IV - In this type, only the extreme posterior portions are joined and the head kidney is not distinguishable. Example - Sea horses and pipefishes (Synganthidae).

Type V - In this last type, the two kidneys are completely separated. Example - Angler (Lophiidae). Ogawa (1961) has reported that all the freshwater teleosts can be grouped into Type I to Type III mentioned above such as - salmon and trout (Salmonidae) are Type-I, carps and minows (Cyprinidae) are of Type II and killfish (Cyprinodontidae), stickleback (Gasterosteidae), sculpin (Cottidae) are of Type III.

Kerr (1919) reported that the posterior kidney of all gnathostomous fishes is developed from the entire caudal portion of the nephrotomic plate. Thus it should be known as ophisthonephros, representing both the embryonic mesonephros and the caudal nephrogenic material that in higher vertebrate forms the metanephros kidney.

Kidney of teleosts is basically polynephric ophisthonephros (Smith, 1960) and its duct is the ophisthonephric (Wolffian) duct formed by the joining of the two mesonephric ducts posteriorly (eg. *Mystus*) or at the point between the kidney and the urinary bladder as in *Labeo* or *Cirrhina*. The urinary bladder may be a simple enlargement of the common mesonephric duct or sac-like structures (*Barbus* and *Mystus*). The bladder opens to the exterior by a common urinogenital pore in male and in female by separate urinary pore.

6.1.2.3. Salt gland or rectal gland

Elasmobranchs have a structure called salt or rectal gland to secret sodium chloride (NaCl) from their bodies. Such fishes require a lower internal NaCl concentration than the surrounding seawater, which causes a concentration gradient favoring the influx of salt. So, they require a way to secrete the salt. This salt gland is present in the rectum of shark which produces a concentrated salt solution for secretion. The thick ascending loop of Henle of kidney and fish gills contain salt secreting cells.

Rectal glands are compound tubular glands (as its boundaries contain tubular cells) with a central canal which continues as a duct. The duct arises from the posterior ventral part of the gland and curves forward before it drains into the intestinal canal posterior to the spiral valve. It has three layers (Conte.1969).

1. An outer capsule containing small arteries and a peripheral connective and muscle tissue layer.
2. A middle glandular layer consisting of a zone of secretory tubules.
3. An inner layer having veins and ducts arranged around a central canal that terminates in a central duct.

There is complex infoldings of the basal and lateral plasmalemma. Mitochondria are numerous in the cells of this gland. Some rough and smooth endoplasmic reticulum and golgi apparatus occur.

These salt-secreting cells transport NaCl by the sodium transport mechanism i.e. the sodium-potassium pump (Na^+/K^+ pump) with the use of the enzyme Na^+/K^+-ATPase without filtration. In the shark, the rectal gland cell cytoplasmic Na^+ is 20mM compared to 280mM in the body fluid. This large Na^+ gradient provides the energy for driving carrier-mediated electroneutral entry of chloride (Cl^-) and sodium (Na^+). Two Cl^- are co-transported with each Na^+ and K^+. The Na^+/

K^+-ATPase activity allows for the movement of NaCl from the blood across the epithelium into the lumen of the salt gland for secretion. Active transport that works against the concentration gradient, produces an increase in the Cl^- concentration in cytoplasm of epithelial cells. This results in the diffusion of ions out of the cell across the apical surface. The build-up of chloride ions at the apical surface attacks Na^+ to diffuse between the cells (the paracellular route).

6.1.2.4. The Chloride Cell

Special columnar flat cells are found in the fish gills, which look like secretory cells. These cells are called chloride cells (Catlett and Millich, 1976) since they were suggested to secrete Cl^- actively from the blood into the seawater. The transport of Cl^- from blood to seawater is in an active electrogenic system, and that Na^+ simply passively follows the gradient created by chloride movement. There is an excellent agreement between the chloride flux and the number of chloride cells, present in the gills. Active Cl^- transport by these cells is the driving force for salt secretion by the gills.

A typical chloride cell's basal portion is towards the blood (internal) and the apical upper part is towards the seawater (Fig 6.6). The basal and lateral sides are extensively infolded to form the cytoplasmic microtubules making the smooth tubular (or microtubular) system (STS). The chloride cells have a large nucleus and feature an abundance of energy-providing mitochondria, which are closely packed with STS leaving a small apical pit. This cell is covered entirely by overlapping epithelial cells except apical region. The developing chloride cells or accessory cells are often found attached with the mature cells but the attachment is leaky. The pavement cells are connected by deep tight junctions.

The STS contains Na^+/K^+-ATPase sodium pump like that of the rectal gland secretory tissue and actively transports Na^+ out of the cell in exchange of K^+ (Karmaky,1986). This pump system functions to maintain high Na^+ gradient with high Na^+ ion in the STS and the adjacent plasma and low Na^+ ion in the chloride cell cytoplasm. This high sodium gradient drives a linked Na^+-Cl^- carrier system, increasing the cytoplasmic Cl^- ion. The accumulation of Cl^- ion in chloride cell increases the cell's electronegativity and Cl^- follows its electrochemical gradient (including a negative to positive transmembrane potential) by passively moving out through the apical pit to the sea via Cl^- ion channel (Karnaky,1986). The Na^+ ion exists passively via the cation selective paracellular pathway ending in leaky junction. Sodium ion thus follows the transepithelial potential gradient created by the movement of Cl^- ion. Utida and Hirano (1973) have reported that both chloride cell number and Na^+/K^+-ATPase activities increase with increasing environmental salinity in case of Japanese eel, *Anguilla japonica*.

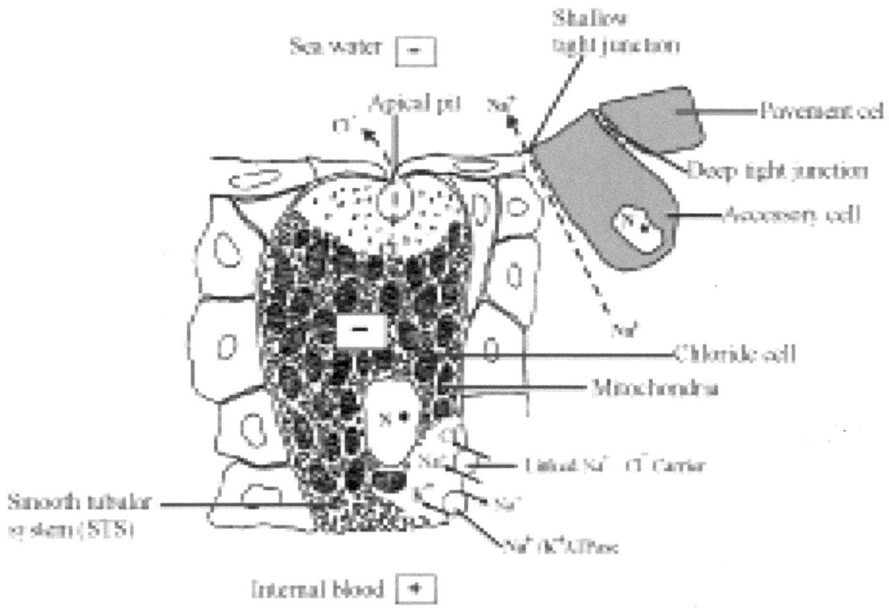

Fig. 6.6 : Structure of chloride cell with the movement of Na^+ and $C1^-$ in sea water fishes.

In freshwater fishes the number of chloride cells is less and these cells take up Ca^{2+} as noticed in Tilapia. Thus, they are termed as calcium cells. In anadromous fish like salmon, these cells multiply in number in the gills as the fish prepare to move to sea.

6.1.3. Functions of Kidney

6.1.3.1. Formation of urine

Nephrons, the functional units of the kidney, produce urine through three stops: glomerular filtration of the blood, selective tubular reabsorption of the glomerular filtrate and tubular secretion of harmful substances (Fig- 6.7).

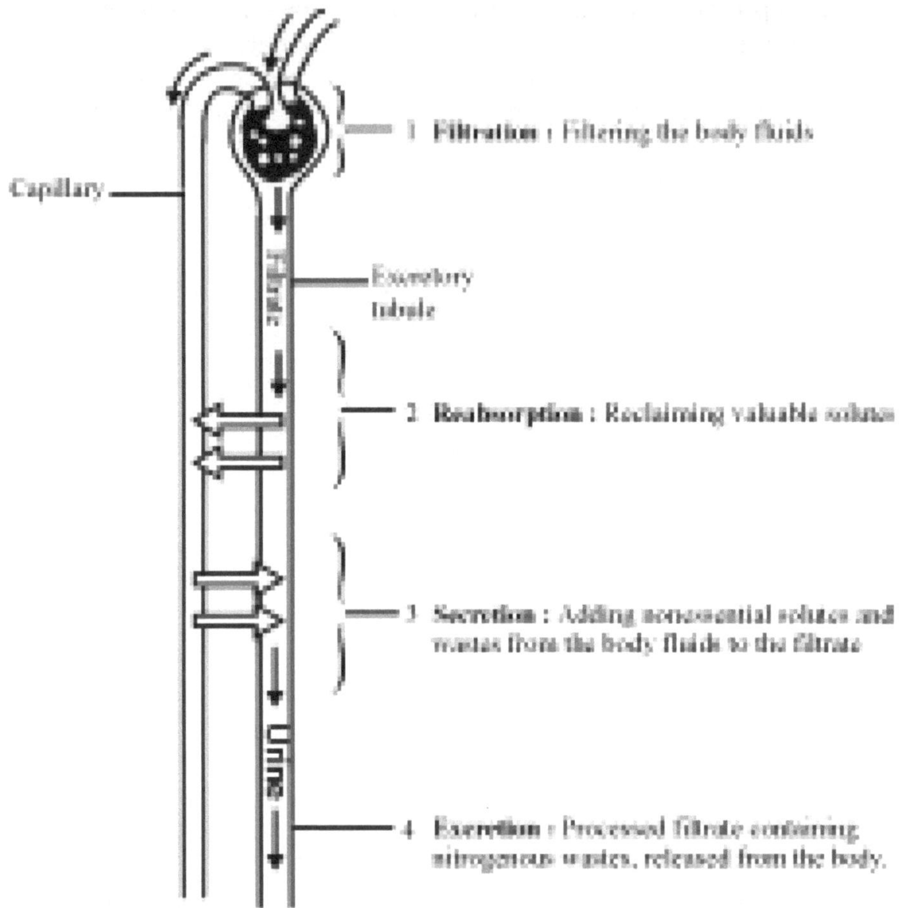

Fig. 6.7 : Steps of urine formation

A. Glomerular filtration

Urine formation begins with filtration of blood through the epithelial walls of the glomerulus and Bowman's capsule. Blood flows from the afferent arteriole into the glomerulus. The ultrafiltration apparatus at the glormerus consists of an ultrafilter made up of a fenestrated capillary endothelium, a fine pored basement membrane and Bowman's capsule. The fluid portion of the blood, which consists of water, metabolic wastes, ions, nutrients and small proteins, is able to move across the capillary wall. Blood cells, larger proteins and other molecules of molecular weight more than 70000d cannot cross and retained in the blood. However, serum albumin with higher molecular weight can pass through the sieve. Materials that enter the glomerulus, leaving the blood, are called glomerular filtrate or primary urine. The ultrafiltration is non-selective and fluids move carrying solute along with. Inorganic ions, small organic molecules, like glucose, urea, amino acids etc., are filtered into the nephric tubules. In the glomerulus, 15 to 25 percent of the plasma's water and solutes are fittered through a single cell layer of the capillary walls, through a basement membrane, and into the lumen of the Bowman's capsule. The filtrate then flows into the renal tubules where reabsorption occurs before urine excretion.

Both kidneys produce glomerular filtrate at the rate of about 125 ml/min or 180 liter/day, equivalent to ten times the blood volume in human. Decrease in glomerular filtration rate (GFR) is a sign of kidney or renal failure. Water and dissolved substances are present in the filtrate at equal concentrations as they are in the blood. So, reabsorption in the renal tubules is essential. About 1250 ml of blood circulates per minute through the two kidneys in human and out of it 650-700ml is plasma. This is called renal plasma flow (RPF). The rate of glomerular filtration or effective filtration pressure (EFP) depends on three factors - the glomerular hydrostatic pressure (GHP) difference between the capillaries and the Bowman capsule (due to blood pressure), the blood colloid osmotic pressure (BCOP), which opposes filtration, and capsular hydrostatic pressure (CHP).

GHP is blood pressure in glomerular capillaries due to narrower efferent arteriole and is the main determinant of EFP. Its value is 75mm Hg. BCOP is osmotic pressure created in the blood of glomerular capillaries due to plasma proteins. Its value is 30mmHg. CHP is caused by the fluid occupying the Bowman's capsule and resist filtration due to the hydraulic permeability of the three-layered tissue separating the capillaries and the lumen of the Bowman's capsule. Its value is 20mmHg. So, EFP = GHP - (BCOP+CHP) i.e. 25mm Hg. Filtration fraction of the plasma passing through kidneys, is filtered at the glomerulus. It is the ratio of GFR to renal plasma flow (RPF).

Glomerular filtration occurs because the pressure of the blood flowing in the glomerular capillaries is higher than the pressure of the filtrate in Bowman's capsule. Blood pressure drives glomerular filtration. Since the process takes advantage of a pressure gradient, glomerular filtration does not require the expenditure of energy by kidney cells. Specialized cells in the nephron wall are able to detect changes in blood pressure and accordingly secrete chemicals that change the diameter of the afferent arterioles connected to the glomerular capillaries. By changing the size of these vessels, the amount of blood flowing through the glomerulus is changed, maintaining a relatively stable rate of glomerular filtration and urine formation. This is auto-regulation of GFR by myogenic mechanism. The other way of GFR regulation is through endocrine responses. Glomerular filtration rate can also be regulated by sympathetic nerve responses that cause the constriction or dilation of the afferent arterioles during stress. Vasoconstriction increases blood pressure and GFR but vasodilation decreases blood pressure.

B. Tubular reabsorption

Both the proximal and distal segments of the renal tubule are responsible for reabsorption of water and solute before the urine passes to the ophisthonephric duct through the collecting duct (Hill, 1976). Reabsorption occurs within the cells of three regions of the nephron and in the collecting duct, even though most reabsorption occurs in proximal tubule. The epithelial cells of the proximal tubule have many microvilli forming brush border, which increases the surface area for reabsorption. Due to tubular reabsorption much of the filtrate passes out of the nephron tubule's epithelial cell and returns to the blood through the peritubular capillaries. About 99% of the water in the original filtrate is reabsorbed and less than 1% of the original NaCl content appears in the final urine. As a result of this reabsorption urine contains mostly waste materials and excess water. In the proximal tubule 70% of the Na^+ ion, water, urea and Cl^- are reabsorbed. Glucose and amino acids are also reabsorbed by active transport. Thus, 75% of the glomerular filtrate is reabsorbed in proximal tubule. Water is reabsorbed in all parts of the tubules except the ascending loop of Henle. Renal threshold of a substance is its highest concentration in the blood up to which it is totally reabsorbed from glomerular filtrate. High threshold substances are almost completely absorbed from the filtrate. Substances such as glucose, amino acids, vitamin C, Na^+ and water have high threshold, thus completely reabsorbed. Glucose has a threshold value of 180mg / 100ml. Low threshold substances like urea, uric acid, xanthine, phosphate are less reabsorbed and non-threshold substances like creatinine, hippuric acid are not reabsorbed.

From the proximal tubule the filtrate passes on to the descending limb of the loop of Henle, the cells of which have no brush border and low permeability to urea and salts besides that of water. As the filtrate descends in the loop of Henle, water diffuses out because of the high salt concentration in the surrounding tissues. So, this part of nephron concentrates the urine. The loop of Henle acts as a counter-current multiplier. For this reason, marine animals have longer loop of Henle to conserve more water by producing more concentrated urine.

From the loop of Henle the filtrate goes to the ascending limb, which is highly permeable to Na^+ and Cl^- ions but impermeable to water and urea. So, the two monovalent ions diffuse out because there is a higher concentration in the filtrate than the surrounding tissues. From the thin segment the filtrate moves to the medullary thick ascending limb that is concerned with the active transport of the monovalent ions out of the lumen into interstitial space. This causes the fluid reaching the distal tubule to be hyposmotic to the interstitial fluid and allows the passive transport of water out of the tubule. The distal tubule transports K^+, H^+ and NH_3 into the lumen, and Na^+, Cl^- and HCO^-_3 out under the influence of hormones and adjusted according to osmotic conditions.

The filtrate goes to the ureters via collecting duct, renal pelvis, and finally released to outside. Under the influence of the pituitary anti-diuretic hormone, the epithelium of the collecting duct is permeable to water. This causes the hyposmotic filtrate to lose water by osmosis and the concentration of salt and urea in urine increases. The urea also helps to draw water out of the filtrate passing down the collecting duct enabling the kidney to excrete urine, which is hypertonic to the body fluid, a property essential for water conservation.

C. Tubular secretion

Substances that are not part of the initial filtrate because of their large size, are actively or passively transported from the blood to the filtrate in different parts of the nephron. Such substances are various toxins, drugs. These are processed in the liver and join with glucouronic acid for removal from blood capillaries in the kidney and transported into the lumen of the nephron. This process of removal is known as tubular secretion. Ions removed from the blood through this process are K^+, H^+, and NH^-_4. The secretion of H^+ is an important way in which kidney helps to control blood H^+.

6.1.3.2. Role of Hormones in Urine Formation

Four major hormones and/or hormone-like substances are related with renal function and help to maintain homeostasis by regulating the concentration and amount of urine to be excreted. These are anti-diuretic hormone (ADH) or vasopressin (pituitary hormone), aldosterone (inter- renal or adrenal cortex hormone), angiotensin II and atrial natriuretic peptide (ANP) or factor (ANF). Water reabsorption in distal tubule is controlled by ADH in negative feedback (Fig-6.8). Water excretion is regulated by increasing permeability of water and salts in the collecting ducts and their transfer is determined by osmotic gradient. When water content in the body is low pituitary secretes ADH, which makes the walls of the tubules and collecting ducts permeable to water. Binding of ADH to receptor molecules leads to a temporary increase in the number of aquaporin proteins in the membranes of collecting duct cells. Water is reabsorbed in the surrounding tissues eliminating hypertonic urine. Decreased blood volume and interstitial fluid level resulting in decreased blood pressure, enhances aldosteron secretion. All Na^+ of the filtrate is reabsorbed by the epithelial cells of the collecting duct in presence of aldosteron in blood and *vice versa*. Retaining Na^+ increases osmotic pressure of blood and reduces water loss from body. In response to low Na^+ (low blood pressure) the juxtaglomerular cells release the enzyme renin, which activates renin-angiotensin-aldosteron system (RAAS) (Fig-6.9) and triggers a series of biochemical reactions, which bring about an increase in GFR. Renin converts angiotensinogen to angiotensin I, which finally is converted to angiotensin II by angiotensin converting enzyme.

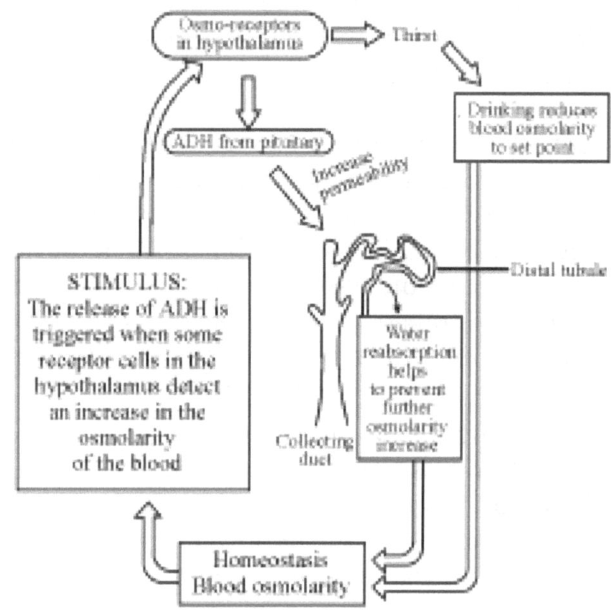

Fig. 6.8 : Role of hormone in urine formation.

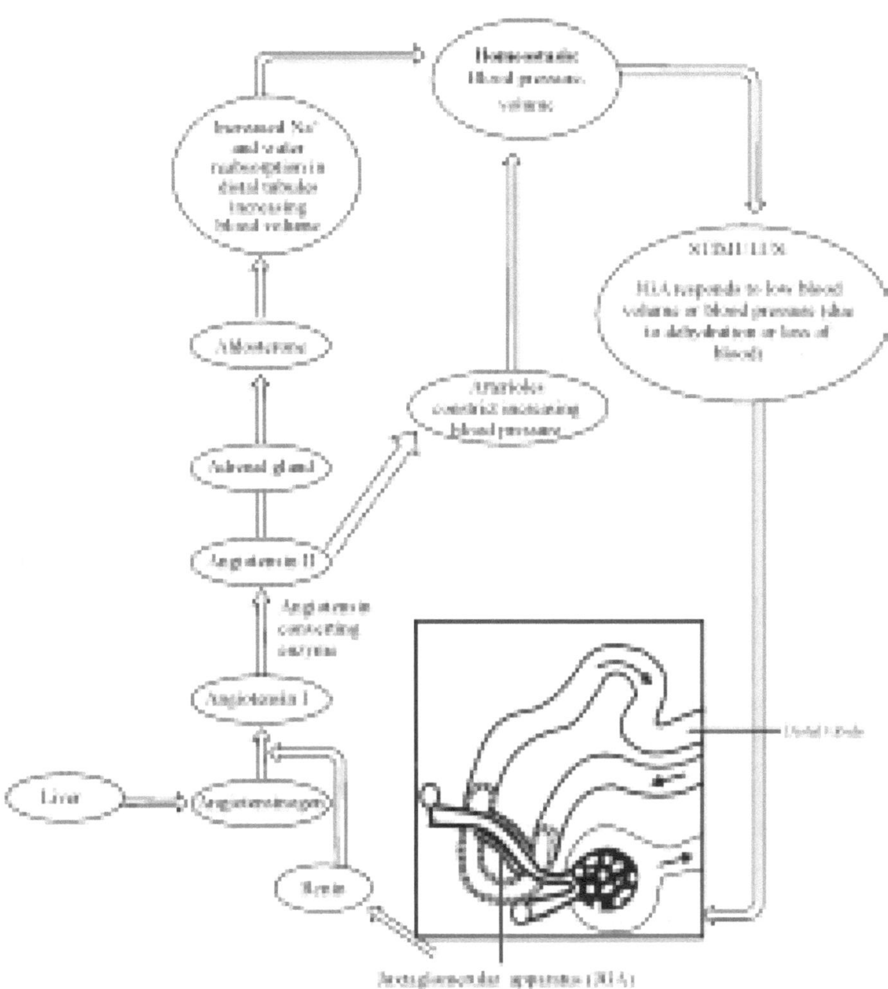

Fig. 6.9 : RAAS system.

The series of reactions includes :

1. Increased synthesis of ADH and aldosteron.
2. Vasoconstriction in arterioles.
3. Stimulation of sodium and water reabsorption.

The ANF opposes the RAAS. The atrial walls release ANF in response to increase blood volume and pressure. ANF inhibits the renin release and Na^+ reabsorption and reduces aldosteron release. So, ANF decreases blood volume and pressure. Thus ADH, RAAS and ANF provide an elaborate system of checks and balances that regulates the kidney's ability to control the osmolarity, salt concentration, blood volume and pressure.

6.1.3.3. Regulation of Body pH by Kidney

Regulation of body pH is done by the carbon dioxide/ bicarbonate(CO_2/HCO_3^-) buffering system in the body in three steps.

1. $CO_2 + H_2O \leftrightarrow H_2CO_3 \leftrightarrow HCO_3^- + H^+$
2. $CO_2 + OH^- + H^+ \leftrightarrow HCO_3^- + H^+$
3. $HOH \leftrightarrow OH^- + H^+$

The excretion of acid by kidney is one of the two major factors which influences this system. The other factor is the removal of CO_2 by lungs. The excretion of H^+ ions (acid) in the urine is mainly responsible for maintaining the plasma HCO_3^- level. The initial glomerular filtrate has HCO_3^- and H^+ concentration high and low, respectively. So, in the process of urine formation, acid must be added to the glomerular filtrate besides the removal of bicarbonate. Thus, the excretion of H^+ and recovery of bicarbonate are both important mechanisms by which the kidney can regulate body pH. The A and B-type special cells of the distal tubule help in this process. The A-type cells are acid-secreting cells, which have a proton ATPase in the apical membrane and a Cl^-/HCO_3^- exchange system in the basolateral membrane. The cells also contain carbonic anhydrase that hydrates CO_2 passing through the membrane to form protons and HCO_3^-. The proton are than pumped back into the lumen and can react with HCO_3^- in the filtrate to form CO_2 and water that can diffuse back into the cells and creat an intake of HCO_3^- back into the blood. The B-type cells are base-secreting cells, which have different form of Cl^-/HCO_3^- exchanger in the apical membrane than the other cell type. These cells secrete HCO_3^- into the lumen of all the tubule in exchange for Cl^- ion.

Another mechanism used in pH regulation is the uptake of H^+ by HPO_3^- and NH_3^+ in the lumen to trap excess H^+ in the filtrate. H^+ ion remains bound so that these protons cannot move back into the epithelial cells and the blood, which could lower the pH.

6.1.4. Osmoregulation in Fish

Fresh-and saline-water are the two different osmotic environments in which fishes remain besides that of the aerial environment for the air-breathers. The aquatic animals are either euryhaline or stenohaline, depending on their ability to tolerate different salinities. Animals whose internal osmotic concentration is the same as the surrounding water are termed as osmoconfermers, whereas those species who maintain an osmotic difference between their body fluid and the external medium are osmoregulaters. Most freshwater and marine fishes are osmoregulators, whereas the invertebrates are osmoconformers. Osmoregulators must spend some energy to maintain the osmotic gradient. The amount of energy required for this purpose differs based on :

1. How different the animal's osmolarity is from its surrounding?
2. How easily water and solutes can move across their body surfaces?
3. The work required to pump solutes across the membrane.

Fishes acquire most of the required water through food, drink, and a smaller amount of water appears due to oxidative metabolism. They lose water through urine, fecal matter. Evaporative loss of water is negligible in fish, even though intake and loss of water across the body surface occur through osmosis. Animals that are protected by covering to stop water loss and gain, have specialized epithelia, which are not water proof and these are exposed to the environment. Examples of these epithelia are at gills, lungs and tracheae. Respiratory surfaces are major avenues for water loss in air-breathing fishes and subjected to dehydration. Thus they drink water to excrete accumulated salts and metabolic waste products.

6.1.4.1. Freshwater fishes

Freshwater fishes are generally hyperosmotic to their external medium. They maintain their water balance by excreting more amount of dilute urine. Salt loss by diffusion are replaced by foods and intake across the gills (Fig-6.10). The problems that they face are that they are subjected to swelling by movement of water into their body owing to the osmotic gradient, and they also have the problem of continual salt loss from the body to the less salt concentrated external water.

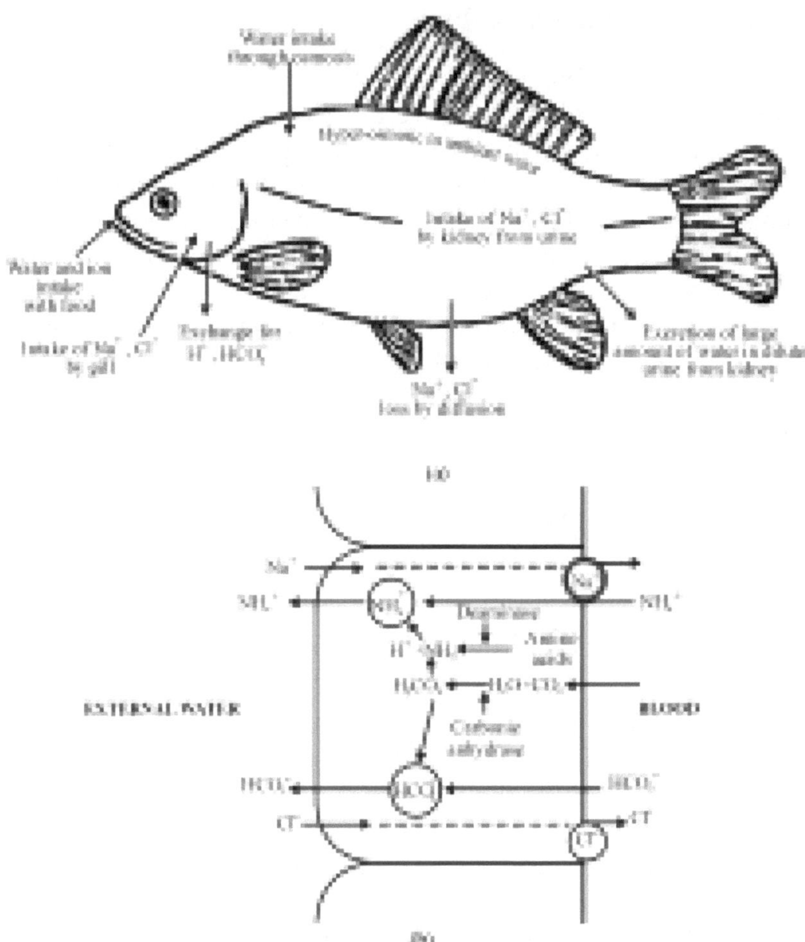

Fig. 6.10 : Osmoregulation (a) and diagrammatic presentation of ionic exchange in branchial cells of freshwater fish (b).

These animals tend to take water passively and remove it through osmotic work of kidney. The kidney absorbs the required salt. A major salt replacement mechanism for freshwater fishes is the active transport of salt from external dilute water across the gill epithelium into the interstitial fluid and blood. Ions like lithium (Li^+), sodium (Na^+), cobalt (CO^{2+}), strontium (Sr^{2+}), calcium (Ca^{2+}), chloride, bromine (Br^-), acid phosphate (HPO_4^+) and sulfate (SO_4^+) are absorbed by gills. Absorption of Cl^- is more compared to others. Hydrogen ion produced from the reaction catalyzed by carbonic anhydrase enzyme, react with NH_3^+ to produce ammonium (NH_4^+). Ammonia is produced due to oxidative deamination of amino acids as nitrogenous waste in the gills but a trace of NH_3^+ will be in the blood due to these oxidative reactions in the liver. Most NH_4^+ diffuses passively from the gills to external water. In exchange, an equivalent amount of Na^+ is carried through the gills by active transport. Bicarbonate (HCO_3^-) from the said enzyme reactions diffuses out of the gill cells and Cl^- are transported actively through gills.

6.1.4.2. Marine fishes

These fishes do not need to spend more energy to regulate the osmolarity of their body fluid. Marine bony fishes are hyposmotic to seawater (their environment is hyperosmotic) and lose water constantly by osmosis and gain salts by both diffusion and from food (Fig-6.11). They compensate by drinking large amount of seawater and pumping out excess salt through gills and skin. Univalent ions, especially the Cl^- pass from the intestine to blood stream and leave through the gills whereas divalent ions like Mg^{2+} and Ca^{2+} remain in the intestine, where the alkaline nature of the fluids promotes their combination with oxides and hydroxides to form insoluble compounds that are then passed with fecal matter. The net result of combined osmotic work of the gills and kidneys in the marine teleosts is a net retention of water. Kidney nephrons in certain marine teleosts have neither glomeruli nor Bowman's capsule. The urine is formed by secretion because there is no specialized mechanism for the production of a filtrate. Chloride cells eliminate excess salt in marine fishes. An important factor in adjustment to lower salinities by marine fishes is the presence of a high concentration of Ca^{2+}. Since Ca^{2+} lowers cell permeability to both salt and water, some marine species remain healthier in a mixture of marine and salt water.

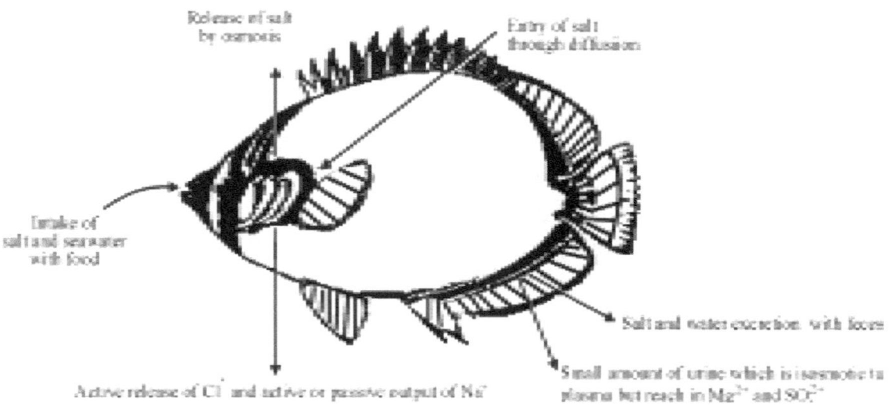

Fig. 6.11 : Osmoregulatin in marine bony fish.

Marine cartilaginous fishes like sharks, rays and skates maintain a lower osmolarity for salt or isosmotic to surrounding seawater. They adjust their internal osmotic pressure that no or little water passes through their permeable membranes (Fig-6.12) and they can sustain osmotic stress. Inorganic salts diffuses through gills, and kidneys and rectal gland excrete salts. The unusually high osmotic concentration of these fishes is maintained by high levels of urea and trimethylamine oxide (TMAO) (less toxic nitrogenous waste) in the blood. Of course high level of urea can damage proteins but the presence of TMAO helps to stabilize these proteins against adverse effects of urea. Little or no urea is lost by the gills of shark.

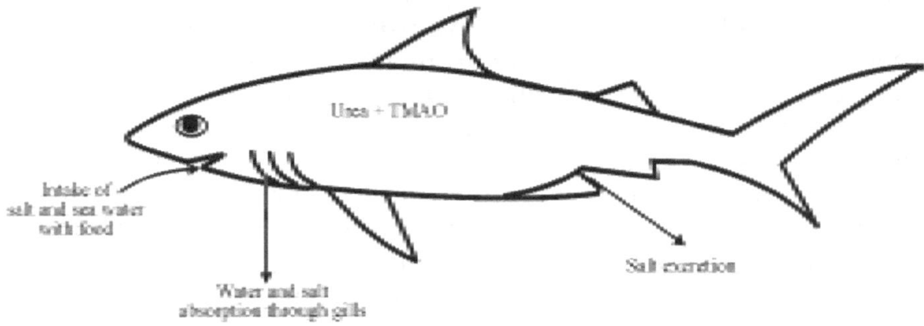

Fig. 6.12 : Osmoregulatin in marine cartilaginous fish.

6.1.4.3. Diadromous Fishes

Diadromous species adjust either to the freshwater or to sea water phase of their life and such adjustments is mainly dependant on genetically determined physiological changes. An ocean fish, like eel, when migrates to freshwater it faces the difficulty of over-hydration and salt depletion, which is *vice versa* to the problem in seawater. The reverse problem is encountered by salmon migrating from freshwater to sea. Thus the migrating species both anadromous and catadromous have well equipped osmotic adaptation. The glomerular kidneys can adjust to differences in urine volumes due to variations in salinities. In addition the gills and oral membranes are capable to adjust both with the intake and the secretion of certain ions against existing diffusion gradients. Chloride cells of marine species can function both as secreting and absorbing sites and must develop well when the fish enters back into the sea after completing freshwater migration.

The case of transition between the two environments depends on anatomical peculiarities such as the gill-to-body surface ratio. Changes in endocrine activities are usually simultaneous or precede with changes in salt and water balance mechanisms. Endocrine glands such as pituitary, thyroid and gonads are mainly concerned with changes in physiological adaptation prior to and during migration. Thyroid activity increases in salmon (*Oncorhynchus*) on down-stream migration, may be to facilitate the energy-demanding process of salt excretion in sea-water, among other functions. Pituitary and gonadal changes often lead to change in appetite, such as the preference for freshwater food and are therefore important in the timing of migration.

6.2. EXCRETION AND OSMOREGULATION IN SHELLFISH

Crustaceans excrete mainly ammonia both in freshwater and marine where water is plentiful besides other substances that are found in small quantities. The minor nitrogenous excretory products include urea, uric acid and some amino acids. The amino acids may be released from the body rather than as having been excreted and their loss as being the result of having bounding surfaces, which must of necessity be permeable to dissolved oxygen, CO_2 and NH_3^+. In crustaceans, part of the nitrogenous waste substances may be deposited in the cuticle and since this is shed at intervals during the growing period, it is possible that some nitrogenous excretion may occur in this manner.

The crustaceans are adapted to many different types of habitat from purely marine to estuarine and from freshwater to an aerial environment. Marine forms have an internal osmotic concentration, which is the same as that of outside, and any dilution of the medium in which they live is closely followed by the dilution of their body fluids to the same level. In consequence they are not very tolerant of the dilution of their surrounding seawater and do not naturally live where such changes occur. The near shore species, on the other hand, can tolerate a range of salinities from pure seawater to near freshwater. Although its internal osmotic concentration does not remain constant, it varies much less when the outside medium is diluted. The tonicity of body fluids of freshwater forms is greater than that of the water outside so that there is a continuous net inward passage of water through the permeable regions of the body like gills. The nitrogenous wastes and excess water are eliminated by the excretory organs.

6.2.1. Excretory organs

Unlike other aquatic animals, the thoracic gills and digestive tract play a role in excretory and osmoregulatory functions besides the green glands of the nephridial system. For clearing of highly soluble waste products like NH_3^+ or in active intake of ions, large gill surfaces provide ready diffusion and transport pathways, given high rates of branchial water exchange and water exchange gradient is enhanced by pleopod-modulated activity (Ahyong *et al.*, 2012). The thick cuticle covering the greater part of the body prevents the entry of water by osmosis into the hemocoel but the thin cuticle over the gills allows water to be drawn continuously into the blood due to the osmotic pressure of the blood as compared with the water flowing over the gills. Crustaceans possess nephrocytes that are mostly present in the hemocoelic spaces of the gill axis and at the base of the legs. These cells are capable of picking up and accumulating waste materials.

Both peristalsis (downstream) and antiperistalsis (upstream) movement of the midgut and hindgut can facilitate the metabolic waste clearance and other releases to or uptake from the external environment. In addition, there is a yet to be explained finding of unique lamellar bodies formed in the walls of the posterior midget caecum of some crustacean species, these being extruded by exocytosis into hemocoel (Felder and Felgenhauer, 1993). It is not clear yet whether these could serve some function in ion or water regulation.

The paired green glands (Fig-6.13) are the main excretory organs. These glands are otherwise known as urinary or nephridial glands. These are highly developed in all decapods. These remain in the hemocoel anterior to the mouth in the head. This gland is a mesodermal structure composed of end or coelomic sac, which is divided into a saccule and labyrinth, tubule (proximal and distal) or nephridial canal and bladder. These nephridial glands are present in the body segment of second antennae and second maxillae with their end sacs arising from an anterior compartment adjacent to those head appendages, antennal or maxillary segment. Based on this, this gland is known as antennal nephridial gland or maxillary nephridial gland, respectively. In most adult crustaceans, only one or the other gland functions. Mostly antennal gland functions in adults. The functional gland may change during the life cycle. However, both types of glands are commonly present at larval stages, the excretory pores or nephridiopores of the antennal glands open on to the underside of the bases of the second antennae and those of maxillary glands open near the bases of the second maxillae.

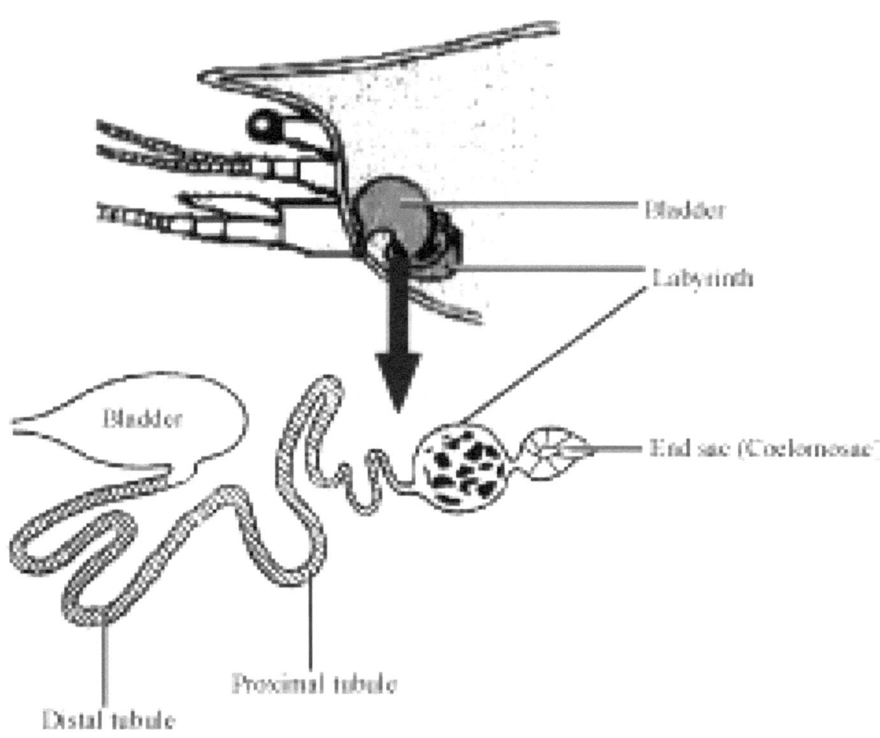

Fig. 6.13 : Maxillary or green gland in decapod.

The mesodermally derived coelomic or end sac, which remains in front of and to both sides of the esophagus, is composed of podocytes, and perform the ultrafiltration function similar to vertebrate glomeruli (Johnson, 1980). Labyrinth walls are greatly folded with an extensive network of coiled cells in the inactive state and columnar cells in the actively secreting state. It is glandular, converting the original vesicle into a spongy mass, and appears to be an important site for reabsorption (Peterson and Loizzi, 1974). The labyrinth cells have a centrally placed nucleus with many mitochondria packed within the extensively infoldings of the basal lamina, characteristic of transport tissue (Mantel and Farmer, 1983). The apical part of these cells has an extensive brush border. The labyrinth is a transport system involved in the movement of ions and reabsorption of proteins (Peterson and Loizzi, 1974). The labyrinth leads to a bladder via the excretory or nephridial tubule. The length of the tubule varies much among the species of decapods. The decapod bladder varies in complexity. It may either be a sample vesicle or give rise to many diverticula that extend to various body parts. The bladder is usually a large reservoir of urine storage and may play an important role in final urine modification

(Riegel, 1972). From the bladder a short duct extends into the basal segment of the second antenna or maxillae, where it opens to the outside on the summit of a small papilla. In brachyurans, the excretory pore is covered by a little movable operculum. No valves or sphincters occur between any of the internal parts of the gland except a valve at the external orifice.

Branches from the antennal and ventral thoracic arteries provide an unusually copious blood supply. Their minute terminal branches, discharge into tiny hemocoelic spaces, which surround all the ramifications of the gland. The bladder receives from the antennal artery a much more restricted arterial blood supply.

In the antennal gland of the lobster certain important differences are noticed. The coelomosac occupies most of the dorsal glandular part, with the labyrinth surrounding its periphery and extending under it ventrally. A tubule is absent, may be because the lobsters secrete scanty, isotonic urine, in contrast to the copious, hypotonic urine secreted by the freshwater crayfish (Brown, 2002). From the anteriomedial portion of the labyrinth a white projection extends venteroposteriorly. This projection contains a number of parallel tubes, which open into the duct from the bladder near its external orifice. There is no direct opening into the bladder, and thus the bladder can be filled only by the backing of urine through its duct. This duct runs downward from the bladder in front of the glandular part of the organ, then turns ventro-posteriorly to the external opening, located in the center of the operculum.

The paired antennary glands of the crabs are on the ventral surface of the head, in a position between the antenna and the eyestalks. Each is partly covered by the stomach, and is pale greenish or yellowish in color. It is composed of tiny coiled tubes. Its structure is almost similar to that in the lobster. A horseshoe-shaped opening at the inner posterior corner of the gland, not often seen, opens upwards from the labyrinth into the bladder. The bladder is very large in size with thin, transparent walls. From the main vesicle of the bladder, located above the gland, a number of lobes extend out in various directions. Most prominent are an epigastric lobe, rising at the front and side of the stomach towards the dorsal surface and a hepatic lobe, which passes posteriorly under the gap of the lateral adductor of the mandible and then turns laterally between the muscle and hepatopancreas. Smaller lobes enter between the hepatopancreas and stomach, and around the esophagus. The cells of all these lobes of the bladder are secretory. The main vesicle of the bladder opens to the exterior by a duct with thin membrane similar to the operculum of crayfish.

6.2.2. Process of Excretion

Even though the complex nephridia like green glands are the excretory organs, most nitrogenous wastes, in the form of NH_4^+ diffuses from the body surface where the exoskeleton is thin as in the gills. These glands appear to function in controlling internal fluid pressure and to some extent, ion content, such as Mg^{2+}. These glands extract nitrogenous waste products and excess water from the blood in a similar manner like the vertebrate kidney. By ultrafiltration, water and dissolved substances from blood pass into the saccule end sac. This end sac excretes compounds of ammonia but other nitrogenous wastes are excreted by other parts of the gland. Some of the more terrestrial forms produce urea or uric acid, which may be stored in special large cells near the bases of the legs or excreted without much loss of water. Uptake of water by the blood increases the filtration pressure within the green gland and the passage of fluid into the saccule. Through the labyrinth and tubule, useful substances return to blood, due to selective reabsorption, whereas the remaining excretory fluid or urine flows into the bladder. Reabsorption requires energy, which is met by cell respiration. The excretory fluid is also collected from the renal sac into the bladder and then excreted out through the renal pores.

The hydrostatic pressure of the colloids in the blood causes water containing crystalloids (including NH_4^+ in the form of ammonium carbonate) to be filtered off from the blood into the end sac then it passes through the green gland to the exterior. Copious amount of urine may be excreted depending on internal and external conditions. The prawn passes out a large volume of urine because the blood is hypertonic to the surrounding medium and water continuously diffuses into blood through gills. Most crustaceans, even many freshwater and terrestrial species, produce urine that is isosmotic with blood. This is not true for of the freshwater crayfish in which the antennal glands produce hyposmotic urine. The urine of crab is isotonic with the blood in conformity with the absence of a tubule in the antennal gland. When in brackish- or fresh-water, the crab excretes copious isotonic urine, and salt loss occurs. The European crab (*Carcinus maenas*) produces urine daily equivalent to 3.6% and one third of its body weight in undiluted seawater (3.4%) and brackishwater (1.41%), respectively (Barnes, 1982). The maintenance of a constant body volume is indicated by the fact that the size increase of a newly molted blue crab (*Callinectes*) is the same regardless of the salinity of the environment. The antennal glands functions to maintain this consistency.

6.2.3. Osmoregulation in Shellfish

Except in crayfish and certain shrimps, the antennal glands do not play a major role in osmotic regulation. For most decapods, the urine is isosmotic with blood even when the animal is in brackishwater.

6.2.3.1. Hemolymph composition and osmoregulation

Hemolymph is the largest tissue in decapods. The make-up and quantity of its constituents change during molt cycle. Hemocyanin is the major constituent of hemolymph. The concentration of hemocyanin is low in postmolt and rises towards premolt (Hagerman, 1983; Mangum *et al.*, 1985). The relative concentrations of free amino acids in crab hemolymph (Yamaoka and Skinner, 1976) and various protein fractions in lobster (Barlow and Ridgway, 1969) and crab (Busselen, 1970) serum vary during the annual cycle. During premolt, glucose (Telford, 1968) and lipid (Spindler-Barth, 1976) levels in hemolymph increase much compared to postmolt. In order to expand the new, soft exoskeleton in postmolt, aquatic decapods must take up water and there is a decrease in glucose that is used as a precursor in chitin synthesis (Hornung and Stevenson, 1971). So, it is expected that the activity of a whole range of physiological processes related to water and ion permeability and regulation vary during molt cycle. A number of ion concentrations in hemolymph are lower in postmolt compared to premolt (Mercaldo-Allen, 1991). Due to the input of water at ecdysis, the osmolarity is lower at postmolt (Mantel *et al.*, 1975). To reduce water loss, urination decreases with a concomitant decrease in NH_3^+ excretion (Regnault, 1979). As a result of this decreased excretory rate, osmoregulatory capacity is lowest just prior to and after ecdysis (Charmantier *et al.*, 1994). Since much of the deposited Ca^{2+} is reabsorbed in preparation for ecdysis, its concentration in the hemolymph increases in premolt (Chang, 1995).

6.2.3.2. Osmoregualtion in different environments

In freshwater, the decapods are osmoregulators, maintaining a minimum blood salt concentration that is hyperosmotic to the surrounding medium while marine species are typical osmoconfermers i.e. the blood salt content is the same as that of the external medium. As in fish, the gills pick up salts from the ventilating current, replacing ions those are lost in the urine and elsewhere. The crayfish antennal gland have a long tubule between the labyrinth and the bladder. Reabsorption of salts by the tubule enables them to produce dilute urine, which helps in osmoregulation. Osmoregulators, which can depend on salt transport by interface epithelia to prevent extracellular disturbance, may have a lower capacity of tissue water regulation when compared with osmoconformers.

According to Florkin (1962) adaptation of euryhaline crustaceans to environmental salinity changes results from two types of mechanisms-

1. Aniosmotic extracellular regulation (AER), which mainly involves ionic exchanges between blood and external medium.
2. Intracellular isosmotic regulation or cell osmotic pressure adaptation in which amino acids play a major role (Schoffeniels and Gilles, 1970).

Thus significant changes in NH_3^+ excretion rate occurs (Spaargaren *et al.*,1982). Still then the salinity effect on nitrogen excretion is dependent on the crustacean species and its osmoregulatory capacity (Regnault, 1984). In other words some species exhibit strong osmoregulatory abilities being capable of hyperosmotic regulation at low salinities as well as hyposmotic regulation at high salinities as in case of the shrimp, *Crangon crangon* (Regnault, 1984). Excretion rate increases over the ebb phase and decreases over the flood phase of the tidal salinity cycle. The euryhaline crustaceans when exposed to a salinity gradient, there is an increase in nitrogen NH_3^+ excretion as salinity decreases and *vice versa*. Coastal species that inhabit areas under tidal influence, thus prone to salinity fluctuations, are necessarily more tolerant to salinity variations, and in general more euryhaline than either marine or freshwater species that may have gone stenohaline (Freire, 2008; Foster *et al.*, 2010).

Intracellular isosmotic regulation mainly resulted from variations in the amino acid concentrations. These variations could be obtained by either synthesis or break down of peptides and proteins (Richard, 1982) or catabolism or synthesis of amino acids (McNamara *et al.*, 2004; Augusto *et al.*, 2007). The second possibility appears better to justify the changes in NH_3^+ excretion with salinity. On the other hand, it was demonstrated that the active Na^+ uptake following the transfer into a hypotonic medium was related to an increased NH_3^+ output (Pressley *et al.*, 1981). A reduction of the Na^+/NH_4^+ exchanges (Pequeux and Gills, 1981) and a possible reversal of these exchanges (Armstrong *et al.*, 1981) were observed following transfer into a hyperosmotic medium.

6.2.3.3. Neuroendocrine regulation of osmoregulation

Gills are the primary organs for salt transport, but in land crabs ion exchanges as well as CO_2 and NH_3^+ excretion are compromised. Urinary salt loss is minimized in land crabs by redirecting the urine across the gills where salt reabsorption occurs. Euryhaline marine crabs utilize apical membrane branchial Na^+/H^+ and Cl^-/HCO_3^- exchange supported by a basal membrane Na^+/K^+-ATPase, but in freshwater crustaceans an apical V-ATPase provides for electrogenic uptake of Cl^- in exchange for HCO_3^- (Morris, 2001). The HCO_3^-, provided by carbonic anhydrase, facilitate CO_2 removal while NH_4^+ can substitute for K^+ in the basal ATPase and for H^+ in the apical exchange. By passing the urine through gills, the NaCl concentration of crabs can be reduced to 5% of that of the hemolymph (Morris, 2001). This provides a filter reabsorption system similar to vertebrate kidney. Crabs have hormonal control over branchial transport processes. Aquatic hyper-regulators release neuroamines from the pericardial organ, including dopamine and 5-hydroxytryptamine (5-HT), which stimulate Na^+/K^+-ATPase activity through cAMP-mediated phosphorylation besides NaCl uptake. Crustacean hyper-glycaemic hormone (CHH) has been proved to have effects on hydromineral regulation (Morris, 2001). In terrestrial crabs there may be controls on both active uptake and diffusive loss. There has been a multiple evolution of a kidney-type system in terrestrial crabs capable of managing salt, CO_2 and NH_3^+ movements.

CHAPTER - 7

Reproductive Physiology

Most of the aquatic organisms spend much of their lives and energies for reproduction. The young ones mature, develop sperms or eggs, spawn, and then recover to repeat the breeding cycle that continues till their death. In a majority of fish species males and females are separate individuals, fertilization is external and large number of eggs are produced (on an annual basis, its fecundity) which develop, hatch and mostly grow without parental care. The sperm and eggs are usually produced in such quantities that they represent a large expenditure of energy. The spawning process may be preceded by migrations or nest building, and followed by care of the young. Thus the success of any fish is ultimately determined by the ability of its members to reproduce successfully in a fluctuating environment and thereby to maintain viable populations. As each fish species occurs under an unique set of ecological conditions, it has an unique reproductive strategy with varied anatomical, physiological, behavioral and bioenergetics adaptations. Unlike the finfishes, in shellfishes sexes are separate. The shellfishes use various signals to attract the opposite sex.

Most mature fishes and crustaceans develop external sexual differences in addition to the primary sexual organs, the gonads. These may help either aquaculturists or prospective mates to recognize the sexes. These are known as secondary sex characters, and may be either accessory to spawn or not. The accessory structures include the intermittent organs of some male crustaceans and fishes, and some fin or mouth structures that may be used in the care of either gaments or youngs. Other sexual difference (mostly common to freshwater fishes) includes differences in size, shape of the body and fins, body color and presence of tubercles. Color difference between the sexes and typical colors for both sexes are most striking among salmonids and many small freshwater tropical fish. Marine species are least sexually differentiated. The sex of the cods, herrings, tunas and mackerels cannot be recognized externally.

The primary sexual characters are the reproductive organs. The secondary sexual characters themselves are of two types - those which have no primary relationship with the reproductive activity and those which are definitely accessory to spawning. A genital papilla marks the male of lampreys (Petromyzonidae), Johnny darters (*Etheostoma nigrum*) and white bass (*Morone chrysops*). The body shape is an important secondary sexual character. Mostly the females are much more pot-bellied than males during breeding season due to the large quantity of sexual products that they contain. Pearl organ or nuptial tubercles appear in males (eg. Cyprinidae, Catostomidae and Osmerus). These tubercles disappear after breeding season under the influence of hormonal secretions. The fins of males are larger than females. In some species the tip of the dorsal fin of the breeding male ends in fleshy knobs. The anal fin of males of several species becomes enlarged into an intermittent copulatory organ.

7.1. REPRODUCTION IN FISHES

Fish reproductive organs include testes and ovaries. In most fishes gonads are paired organs of similar size, which can be partially or totally fused. The secondary sex organs increase reproductive fitness. Fishes are bisexual. Some species are hermaphrodite and even parthenogenic reproduction occurs in some cases. In the most common bisexual reproduction, sperms and eggs develop in separate males and females. In hermaphroditism (intersexuality) of Sparidae and Serranidae, the eggs and sperms develop in one individual having both sexes and self-fertilization takes place. Hermaphroditic sex glands are also found in several fish species, including some trouts (salmonoids), perches (*Perca*), walleyes (*Stizostedion*), darters (*Etheostoma*). Some seabasses are protandric hermaphrodites, being male at first and female later. In parthenogenesis or gynogenesis the sperms simply induce the egg to develop the young without fertilization. This is a condition that occurs in live-bearing Amazon molly (*Poecilia formosa*) and is also known as Poeciliopsis.

7.1.1. Reproductive System

The symmetrical gonads develop from the coelomic epithelium, and are suspended by mesenteries across the roof of the body cavity in close association with the kidneys. The mesenteries in female is called mesovarium and in male it known as mesorchium. In elasmobranches, gonads are associated with hemopoietic epigonal organ. They extend upto cloaca. Epigonal organ may be relatively small as in *Heptanchus* or absent as in *Squalus*. In *Scyllium*, the testis is fused with these organs. It is paired in *Cetorhinus* where the right one is slightly dorsal to the ovary and suspended by a posterior extension of mesovarium. The anterior portion is fused with the posterior portion of the ovary. But the left one is not associated with ovary.

7.1.1.1. Condition factor

A positive relationship exists between length and weight but the extent of this relationship varies with species and there is no effect of season or sex. Based on the length and weight, the condition factor is calculated using the formula :

$$K = \frac{W}{w}$$

Where K is the relative condition factor, W is the observed weight of the fish and w is the calculated weight of the fish. This factor can be used to express the relative health of the fish. The values of K differ with the season, and are influenced by maturity of gonads and spawning, being maximum during spawning season.

7.1.1.2. Male reproductive organs

The paired, elongated and flattened testes are situated on either side ventral to kidneys in the posterior part of the abdominal cavity. The testes are attached to body wall and air-bladder by mesorchia, which is vascularized and contain nerve fibers. In dipnoi (*Neoceratodus* and *Protopterus*), the right testis is attached anteriorly to the tip of the liver while the left one extends forward to the region of ductus Cuvieri. The right one is usually larger than left. The testes have two major functions, the production of spermatozoa (spermatogenesis) and another function is the production of steroid hormones (steroidogenesis). In some fishes, the anterior three fourth part of each testis is functional and the rest posterior part is sterile as in *Mystus seenghala* and *Barbus tor*. The functional and sterile part of the testis is structurally and functionally different. The posterior sterile portion contains empty lamellae, probably for the storage of the sperms during breading season. The size of testes increases during breeding season. But in *Rita rita* and *Mystus*, the posterior part of the testis is glandular. In mature gobies, *Acanthogobius fluviatilis* the testes are small and thread like. In Syngnathids, the testis is simple tubes. The testes of teleosts are long and rounded. In some sharks, the testes are associated with an epigonal organ, which is lymphoid in nature. In fishes like *Rita* and *Glyptothorax*, the testes bear a large number of lobules, which during the breeding season appear as finger like projections.

Paired glandular seminal vesicles are present as outgrowths of the hinder ends of the vas deferentia in some teleosts like *Clarias batrachus, C. lazera, Heteropneustes* fosilis. These are secretory in nature and show periodical changes in correlation with the testicular cycle. The fluid secreted from these vesicles probably serves to keep the sperms in active and viable condition and/or nourishes the sperms. The two sperm ducts join posteriorly and open into the urinogenital papilla.

Histologically the testis is made up off many seminiferous tubules or lobules, which are separated from each other by a thin connective tissue, stroma (Fig-7.1). The walls of the lobules are not lined by a permanent germinal epithelium. The tunica propria of the connective tissue projected into the lumen forming tubes. The blind end is the site of primary spermatocytes. The lobule has interstitial and lobular parts. The former part consists of interstitial (Leydig) cells, fibroblasts, and blood and lymph vessels. The other part contains germ and somatic cells. The somatic cells are positive for lipid and cholesterol, and are similar to mammalian leydig cells (O'Halloran and Idler, 1970). The lobules open into a spermatic duct which is generally lined by a secretory epithelium.

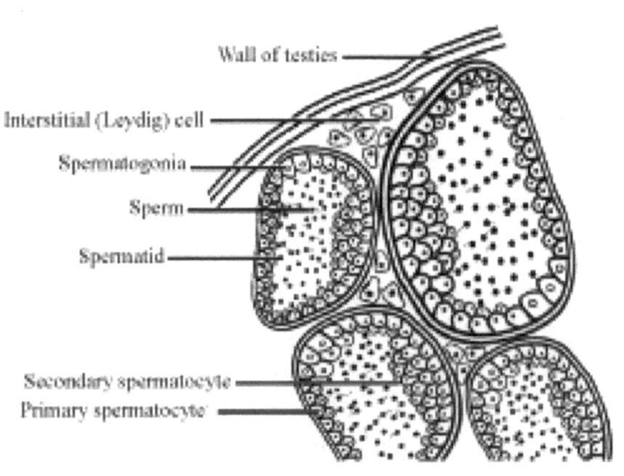

Fig. 7.1 : Histology of fish mature testis

7.1.1.3. Female reproductive organs

The female reproductive system consists of ovaries, oviduct, and in some fishes pseudocopulatory papilla. Fish ovaries may be of three types - gymnovarian, secondary gymnovarian or cystovarian. In gymnovarian, the oocytes are released directly into the coelomic cavity and then enter the ostium. Then, through the oviduct they are released.

Secondary gymnovarian ovaries release ova into the coelom from which they go directly into the oviduct. Both the gymnovaries are primitive type found in lungfish, sturgeon, bowfin salmonids and a few other teleosts. Cystovaries characterize most teleosts where the ovary lumen has continuity with the oviduct and both oviducts join each other to open out through genital pore. In this type of ovary, the mature oocytes are released by pore to the outside.

The ovaries are elongated sac-like structures. The anterior ends of the two are free but mostly fused posteriorly. The posterior part of each ovary develops a short oviduct that unites with each other and open to outside by a separate genital aperture or by a common urinogenital opening. In eel the oviducts are absent and the ova are discharged by genital pore (Parkar and Haswell, 1967). Immature ovaries are thin, flaccid and pinkish or reddish in color. But on maturity they become lobulated, enlarged, and yellowish in color due to the presence of ripe ova.

7.1.1.4. Size at first maturity

Size and age at first sexual maturity of fish species are variable depending on the environmental factors, particularly temperature. Indian major carps and freshwater catfishes attain sexual maturity in the second year and are able to breed up to five years of age. The males mature earlier than females. The female carps grow faster than males but reverse for catfishes. For the purpose of breeding three year old catfish of weight 1.4kg is preferred. Tilapia attains first sexual maturity at 4 to 5 months of age when it is 10 to 17cm length. The age at first maturity of *Mugil cephalus* varies between 2 to 4 years of age depending on the environmental factors and food availability. The milk fish (*Chanos chanos*) was reported to attain sexual maturity in both nature and floating cages in 3 to 5 years. However, in ponds and concrete tanks it matures within 8 to 10 years (Bagarinao, 1994). Trouts attain maturity at the end of second year. Female trouts of 3 years and above are usually used for breeding. As the age increases, fecundity also increases (Huet, 1986).

7.1.1.5. Spermatogenesis or Spermiogenesis

It is the developmental process through which the spermatogonic cells are transformed from undifferentiated diploid (2n) spermatogonia into highly specialized haploid (n) spermatozoa. There are three principal phases :

1. The spermatogonial phase or spermatocytogenesis.
2. The meiotic phase or meiosis.
3. Spermatid phase or spermiogenesis.

The spermatogonia are large oval cells with a round centrally placed nucleus and clear nucleolus. These cells multiply and give rise to primary spermatocytes that are smaller in size (Fig-7.2a). These undergo various stages of division during which the chromatin matter is visible. The primary spermatocytes then become secondary spermatocyte by maturation (or meiosis) division.

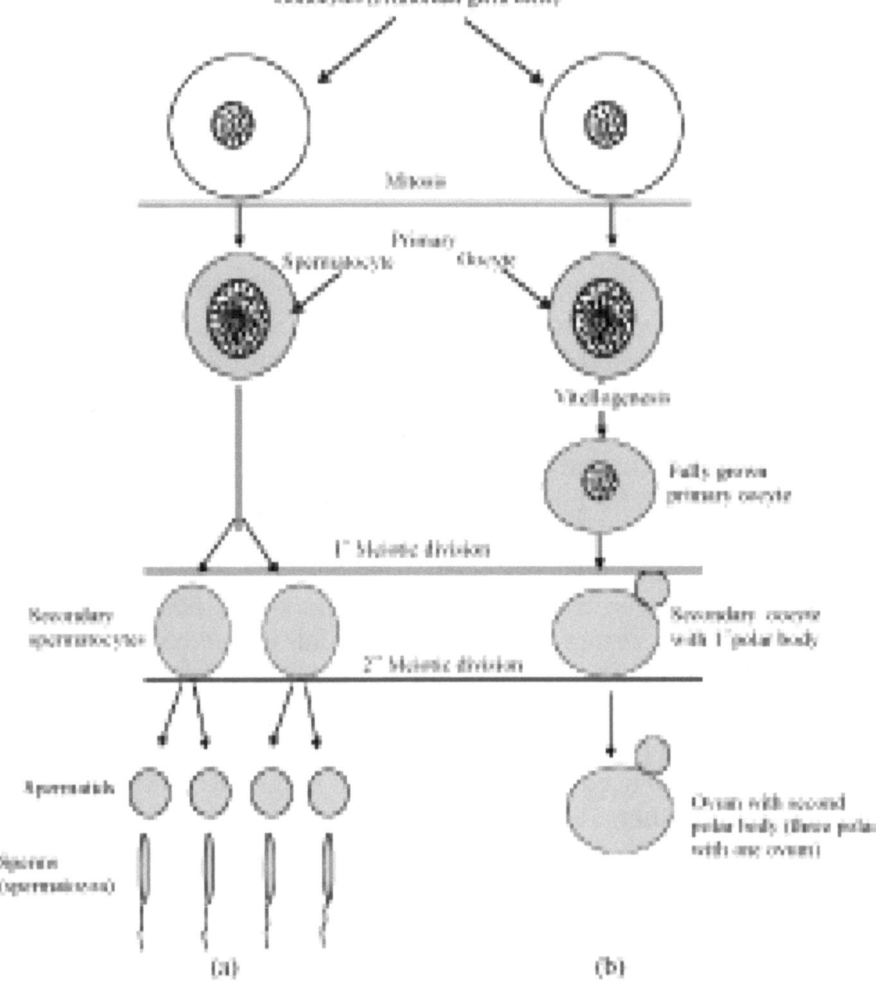

Fig. 7.2 : Diagrammatic presentation of spermatogenesis (a) and oogenesis (b)

The secondary spermatocytes are similar in size with primary ones but their chromatin material is seen in the form of a thick clump. These secondary spermatocytes divide to give rise to spermatids, which finally develop to sperms after reduction in size. The spermatozoa or sperms of different fishes contain different hardy materials. They differ in shape also (Fig-7.3). In the process of spermatogenesis the sperm cells develop a tail that makes them motile, and with which, they make their way into the eggs for fertilization. Large numbers of sperms are produced to ensure fertilization. The spermatozoa and the secretions of the sperm ducts compose the milt that the male fish exudes during spawning. The sperms are inactive and immobile until the secretions occur. The life-span of sperms varies according to the species and substrate into which they are deposited. In water, they have much shorter life than in the body fluid of female. However, if the water has nearly the same salt content as the fish body fluid, the sperm cells live longer. They live longer at low temperatures.

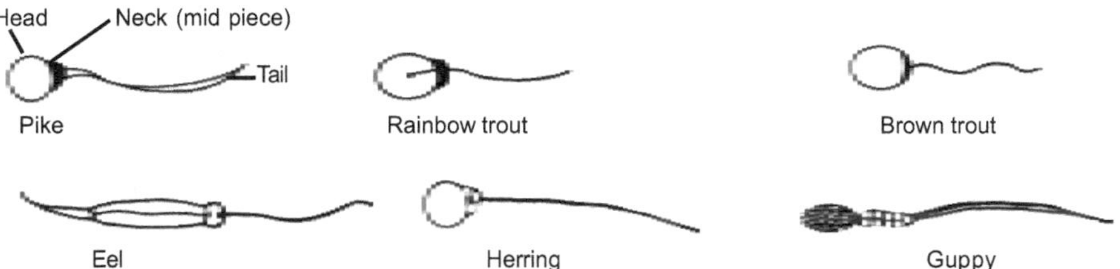

Fig. 7.3 : Enlarged spermatozoa of different fish species showing variations.

7.1.1.6. Oogenesis

The ovarian wall becomes thin and richly vascular during breeding season. The wall has three layers - the outermost thin peritoneum, the middle thicker tunica albuginea made up of connective tissue, muscle fibers and blood capillaries, and innermost germinal epithelium that projects inwardly into ovacoel and gives rise to ovigerous lamellae (Fig-7.4). These lamellae are the place for the development of oocytes and contain large number of oogonia or germ calls in clusters. The oogonia develop from the germinal epithelium. The ovary contains supporting tissue knows as stroma that contains oogonia and oocytes. An oogonium has a large nucleus, a thin layer of ooplasm. The chromosome becomes thread-like in the nucleus (leptotene stage). The chromosome undergoes zygotene stage followed by pachytene and diptotene stage. At this stage of development, oocyte is surrounded by a single layer of follicular cells in cyclostomes and teleosts but in elasmobranches it is multilayered. The egg or vitelline envelop is formed by following the main layers consisting of a cellular plasma membrane (oolemma), acellular zona radiata (zona pellucida or egg shell) and cellular layers consisting granulosa cells, basement lamina and outermost theca cells.

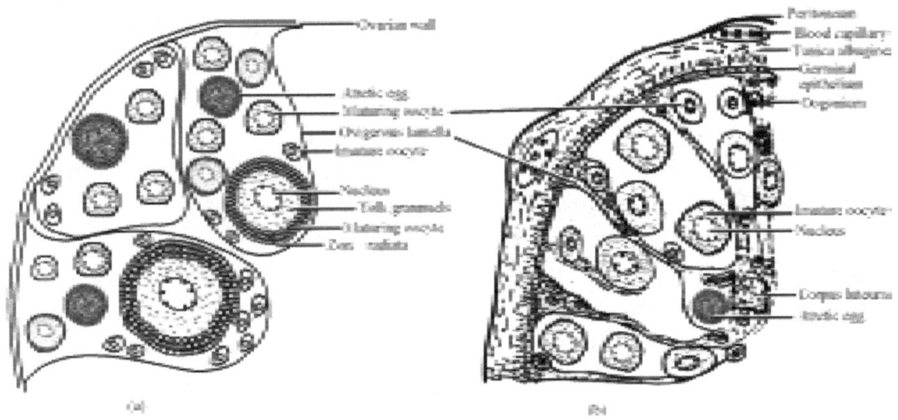

Fig. 7.4 : Histology of fish mature ovary (a) and ovary during non-breeding season (b).

Each oogonium passes through a number of maturation stages to become an ovum. The process of development is known as oogenesis (Fig-7.2b). Changes in the nucleus, ooplasm and the surrounding layers characterize the oocyte maturation process. The developing eggs are supplied with yolk by the follicular cells during vitellogenesis. At the completion of maturation, the oocyte becomes free of the follicle during ovulation and water is taken up causing the egg to swell.

After oocyte release, post-ovulatory follicles are formed. They do not have endocrine function. They have a wide irregular lumen, and are rapidly reabsorbed in a process involving the adaptosis of follicular cells. A degenerative process, called follicular atresia, reabsorbs vitellogenic oocytes that are not spawned (Guraya, 1993). This process can also occurs, but less frequently, in oocytes in other development stages.

7.1.1.7. Vitellogenesis

It is the process of incorporation of the vitellogenin protein into the oocyte and their processing to form yolk. Thus, the ooplasm of the oocyte is filled with yolk granules leaving a small portion of ooplasm with germinal vesicle at the animal pole. Lipids and vitamins are also incorporated into the oocytes. The basophilic ooplasm becomes acidophilic due to vitellogenesis. There are three yolk substances, the yolk vesicles, yolk granules and oil droplets. The yolk vesicles are provided with glycoproteins. The yolk vesicles later become the cortical alveoli to take part in the formation of perivitelline space. By the time of fertilization, the oocytes contain mRNA, proteins, lipids, carbohydrates, vitamins and hormones, which are necessary for the development of the embryo.

Blood of mature oviparous and ovoviviparous females contains calcium binding phospholipoglycoprotein, vitellogenin that are synthesized in liver under the influence of the female sex steroid hormones, particularly 17 -β-estradiol. Vitellogenin passes from the plasma of the female to the oocytes in which it gives rise to the yolk protein lipovitellin and phosvitin (Wallace, 1978). The vitellogenin passes from theca capillaries to granulosa layer to arrive at the oocyte surface through the pore canals of the zona radiate. Then the vitellogenin is sequestered by receptor-mediated endocytes, involving specific receptors in the endocytotic clathrin-coated pits of vesicles. The coated vesicles move into peripheral ooplasm and fuse with lysosomes forming multivesicular bodies. Within these bodies, vitellogenin are cleaved by lysosomal enzyme, cathepsin-D to form small yolk proteins. Completion of vitellogenesis occurs after the movement of germinal vesicles, fusion of yolk granules and grouping of oil droplets. Craik (1978) measured the rate at which vitellogenin is synthesized and converted into yolk granules. At least three types of vitellogenin have been found in the oocytes of teleosts. They are vitellogenin A, B and C.

After the first and second meiotic division, the division is arrested at metaphase and the egg is ovulated. Chieffi (1962) reported that true corpora lutea are found in both oviparous and viviparous elasmobranchs. In *S. stellaris* (oviparous), they are assumed to develop from ovulated follicles and known as corpora lutea and in *I. marmorata* (viviparous) by follicular atresia and known as corpora atretica.

7.1.1.8. Ovulation

In the females, the eggs are formed in two ovaries (sometimes only in one) and pass to outside through genital pore. During ovulation, granulosa, basement lamina and thecal layers are shed and ripe ova have tough zona radiate above the oolemma. The egg-shell is involved in fertilization. After fertilization, the egg envelop, also named as chorion at this stage, protects embryo in the aquatic environment. Over 97% of all known fishes are oviparous, that is the eggs are fertilized outside as the fish shed their gaments into the surrounding water. Examples of oviparous fish include salmon, goldfish, carp, tuna, eels etc. A few oviparous fish practice internal fertilization with the male using the interomittent organ to deliver sperms into the genital opening of the female. This is noticed in oviparous sharks (such as horn shark) and rays (such as skates). In these fishes, the male is equipped with a pair of modified pelvic fins, known as claspers.

In some teleosts the urinogenital papilla or anal fin is enlarged and modified for the transfer of sperms for internal fertilization. In *Gambusia,* the anal fin rays of male are modified to form an elongated copulatory organ that is known as gonopodium. In *Apogon imberbis,* the female has an elongated genital papilla which is introduced into the male for receiving the sperm (Garnaud, 1950). The eggs are fertilized internally but are shed before development. Marine fishes produce more number of large eggs that are mostly released into the open water. The eggs have an average diameter of 1 millimeter (0.039 inch).

Many bony fishes (teleosts), elasmobranches (skates, sharks and rays) bear live young. Such fishes have evolved various types of internal incubation or gestation. Some varieties of cichlids and catfishes incubate eggs in their mouth. In seahorse (*Hippocampus*) and pipefish (*Syngnathus*), the females place egg in the brood pouch of the male (Lagler *et al.,* 1977). Marine catfishes (Ariidae) use the mouth as an oral incubator. Among live-bearing bony fishes (Osteichthyes) development takes place within ovaries. In some cases, embryo development is intra-follicular. Here the egg follicle is vascularized, dilated and fluid-filled. The embryo without a chorion envelop, develops within the original egg follicle, ovulation and birth are simultaneous. In other cases, the egg may be fertilized while still in its follicle but completes development after leaving the ovigerous follicle within the cavity of the ovary. In all livebearers, the youngs are born at a relatively large size and are few in number.

In some bony fishes the eggs develop within the female after internal fertilization but receive little or no nourishment directly from the mother and depend on yolk. The embryo develops in its own egg case and the young ones emerge when the egg hatch. This is known as ovoviviparous. Familiar examples of fishes are guppies, angel sharks and coelacanths. Some species of fish are viviparous. In these species, the female retains the eggs and nourishes the embryos. The viviparous species have structure similar to the placenta of mammals. Surf perches, splitfins and limon shark are viviparous. Some viviparous fish exhibit oophagy, in which the developing embryos eat other eggs produced by the mother. This has been observed mainly in sharks, such as shortfin mako, porbeagle and in bony fish such as halfback (*Nomorhamphus ebrardtii*). Intra-uterine cannibalism is an even more unusual mode of vivipary, in which the largest embryo eats smaller and weaker ones. This behavior is most commonly found in grey nurse shark and halfbeak. Among the rays, skates and Rajiformes, ovoviviparity is the rule, with the oviparous skates (Rajidae) being the exception. The oviparous sharks and related forms are bottom dwellers of relatively shallow waters and are never large, whereas the live-bearing ones range from small to vary big, and are widely variable in their habits and distribution. Ovoviviparous and viviparous fish are commonly referred as live bearers.

Successful reproduction and in many cases defense of the young are assured by stereotypical but often elaborate courtship and parental behaviour, either by the male or the female or both. The oviparous fish species prepare nests by hollowing out depressions in the sand bottom (example cichlids), build nests with plant materials and sticky threads excreted by the kidneys (example sticklebacks) or blow a cluster of mucus-covered bubbles at the water surfaces (example gouramis). The eggs are laid in these structures. Some fishes, like salmon, undergo long migrations from the ocean to large rivers to spawn in the gravel beds where they themselves hatch (anadromous fishes). Freshwater eels (Anguillidae) migrate to sea to spawn (catadromous fishes). Other fishes undertake shorter migrations from lakes to streams, within the ocean or enter spawning habitats that they do not ordinarily occupy in other ways.

7.1.1.9. Fecundity

The number of eggs laid by a female of a particular species during breeding season is termed as fecundity. So, it is the reproductive capacity of that species. The fecundity varies with size and age of the fish species and environmental condition. The fecundity can be measured by two ways such as :

1. **Volumetric method -** In this method the total volume of the ovary is measured and the numbers of ova in the ovary are counted.
2. **Gravimetric method -** In this method preserved ovary is used. After determining the weight of the ovary, the number of ova in three small ovary pieces of 100 mg each is counted.

The total number of ova are then calculated as $F = \frac{S \times OW}{100}$

where F is the fecundity, S is the average number of ova obtained from three samples of 100 mg each and OW is the total ovary weight.

Fecundity of *Cirrhina mrigala* varies from 75,900 to 11, 23,200 for the fish size of 349 to 810 mm length (Hanumantha Rao, 1974). This ranges from 47,168 to 3, 80,714 in *Labeo rohita* of size 270-490 mm length (Joshi and Khanna, 1981). Fecundity of fish is directly related to the fish length and weight. Fecundity is also related with the diet of the brood fish. Increase level of fat content of the brood fish diet reduces fecundity. In general, fecundity increases with the size of the female and the relationship is: $F= aL^b$ where F is the fecundity, L is the fish length, and a and b are constants derived from the data (Bagenal, 1978).

Larger fishes produce relatively more eggs as per the above relationship. Larger fishes often give large size eggs. While the above relationship seems to hold true for many species, there are many factors which complicate the interpretation of fecundity data, mostly in relation to the investment of energy. Some of the complications (Bagenal, 1978) are :

1. The relationship between fecundity and fertility.
2. The fecundities of multiple spawners.
3. The fecundities of viviparous species and those with parental care.
4. The relationship between fecundity and egg size.
5. The relationship between population density and fecundity.
6. The impact of environmental factors on fecundity.

Survivorship rates are inversely related to fecundity as fishes with high fecundity have high death rates, especially through the free embryo and larval stages. The frequency of reproduction is another characteristic of fish life, which reflects the predictability of the environment in which they live. The two basic strategies here are semelparity and iteroparity. The semelparity is the big bang reproduction where the adults spawn and die (example Pacific salmon). The iteroparity is repeated reproduction (example most seasonal breeders). If conditions are favorable, a semelparity fish may have very high reproductive success, otherwise it may lose out completely. Thus, most fishes increase their fitness by adapting a bet-hedging strategy of not putting all their energy into one spawn (Murphy, 1968).

7.1.2. Physiological Adaptation for Reproduction

The knowledge of the reproductive physiology and the adaptations in fish is essential prerequisite for making successful seed production of fish and viable on commercial scale. The seasonal breeders exhibit circamnual rhythm in gonadal development. The reproductive cycle of fishes are closely tied to environmental changes, particularly seasonal changes in light and temperature. These two important environmental factors act directly or through sense organs or glands that secrete hormones for reproduction, which in turn produce the appropriate physiological or behavioral responses for reproductive cycles. Higher temperature inhibits spawning regardless of photoperiod because they inhibit the transformation of spermatogonia to spermatocytes in male (spermatogenesis) and inhibit the formation of the vitelline membrane in females at the final stage in vitellogenesis. Water chemistry can also affect maturities; low pH can create severe problems in the gonadal maturation (Weimer *et al.*, 1986).

In India, generally fish gonads undergo resting phase from October - November to February, preparatory phase from latter part of February to March, sexually active phase from April to August and an inhibitory phase from August - September to October - November. During the above phases many physiological and developmental changes occur in fish. The ovarian developments consist of multiplication of oogonial cells to produce oocytes, growth of oocytes and vitellogenesis. There is also an increase in the hydration and vascularisation of the ovary.

Oocyte hydration provides water cushion to embryos to survive in the hyperosmotic sea water. It also helps to maintain buoyancy of eggs, increase their survival and dispersal in the ocean. The pelagophil species produce highly hydrated, floating eggs than benthophil species. Amino acids, produced by the hydrolysis of yolk protein, increase the oocyte osmotic pressure and allow water uptake. The free amino acids also play an important role as osmotic effectors during oocyte hydration. In pelagophil species, K^+ ions have also been related to increased oocyte osmolality during maturation, besides the increase of Mg^{2+}, Ca^{2+}, Cl^- total ammonium (NH_4^+) and/or inorganic phosphorus in the maturation cycle. Discovery of molecular water channels, aquaporins, has prompted detailed investigations into molecular mechanisms involved in oocyte hydration of marine teleosts (Pal *et al.*, 2011). Aquaporin is translocated transiently into oocyte plasma membrane during oocyte maturation, immediately after germinal vesicle break down and before complete hydrolysis of yolk proteins. Accumulation of osmotic effectors, aquaporin 1b and intracellular trafficking are two highly regulated mechanisms that allow water influx into oocyte (Pal *et al.*, 2011).

7.1.3. Endocrine Regulation of Reproduction

Internally, the gonadal maturation processes are regulated by the secretions of gonadotropin releasing hormone (GnRH) by the hypothalamus and gonadotropic hormones or gonadotropins (follicle stimulating hormone or FSH or GtH-I and luteinizing hormone or LH or GtH-II) by the anterior pituitary or hypophysis. Pituitary hormones play central role in regulating vitellogenesis, gametogenesis and secretion of gonadal steroid hormones for the development of sexual behavior and secondary sexual characters. The characters of the hormone have been described in detail by Samantaray (2012). In fish, gonadotropin binds to its receptors in the ovarian (theca and granulose) and testicular (Leydig cells) somatic cells that initiate signal transduction cascade resulting in the formation of maturation-inducing steroids (MIS), 17α, 20β - dihydroxy-4-prognen-3-one (DHP) or 17α, 20β, 21-trihydroxy-4-pregnen-3-one (20β-S), with maturation promoting factor (MPF), a complex of dimeric protein of two monomers, cyclin B and cdc-2-kinase. MPF affects oocyte germinal vesicle break down and final maturation.

Increasing temperature and photoperiodicity induce central nervous system for GnRH secretion. A number of monoamines have been implicated in the synthesis and secretion of GnRH. For example dopaminergic neurons are shown to exert inhibitory effect on GnRH neuron. The GnRH stimulates pituitary gonadotroph cells to produce gonadotropins. Many brain hormones and neurotransmitters and amino acids are involved in the release of gonadotropins. They are GnRH, neuropeptide Y (NPY), γ-aminobutyric acid (GABA), taurine, glutamate, aspartate, dopamine, norepinephrine, bombesin, cholocystokinin, galanin, activin/inhibin, nicotine and serotonin. In addition, cholinergic nerves may also participate in this regulation. The GnRH exerts its regulatory role through recognition and binding by specific membrane associated receptors belonging to members of rhodopsin - like G-protein coupled receptor family.

GtH-I is secreted during vitellogenic phase of oocyte development and GtH-II is released during post-vitellogenic phase. GtH-I induces the secretion of estradiol from follicle cells. The thecal and granulosa cells have receptors for GtH-I. The GtH-I stimulates the thecal cells to produce testosteron from the cholesterol precursor (Samantaray, 2012). From cholesterol pregnenolone is produced, which is further converted in step-wise manner to progesterone, 17αOH-progesteron, androstenedion and testosteron with the use of various enzymes. The testosterone so produced in the thecal cells enters into the granulosa cells, which also possess GtH-I receptor. Binding of GtH-I to the receptor of granulosa cells induces the conversion of testosteron to estradiol by an aromatase enzyme. The estradiol is liberated to blood circulation and acts on the liver cells to produce vitellogenin. Simultaneously growth hormone (another pituitary hormone) potentiates effects of estradiol in stimulating vitellogenin synthesis.

Major endocrine events associated with termination of vitellogenesis and resumption of meiosis (oocyte maturation) show steady increase in plasma GtH-II and GtH-II related ovarian follicle steroidogenesis. In a post-vitellogenic oocyte, the nucleus or germinal vesicle, which is in pachytene stage of prophase of first meiotic division, undergoes further steps in meiosis under the influence of GtH-II. GtH-II induces the follicle cells of oocytes to produce a hormone called maturation inducing hormone (MIH), which is 17α, 20β-dihydroxy-4-pregnene-3-one from 17α - OH progesterone. The MIH is species-specific derivative of progesterone, which binds to oocyte membrane-specific receptors to activate MPF in ooplasm that finally triggers dissolution of germinal vesicle and reinitiates meiosis.

Leydig cell steroidogenesis is directly regulated by GtH-I and II. Functions of sertoli cells are structural, nutritional and regulational (paracrine) support for germ cell development and are regulated by GtH-I. Progestagens, androgens and estrogens are mainly produced by testes. Androgens such as testosterone and 11-ketotestosteron are effective in supporting either partial or whole process of spermatogenesis. Progestins induce spermiation in Salmonidae and Cyprinidae.

Sex attraction, synchronization of reproductive processes breeding migration and parent-offspring recognition are done by the semiochemicals, the pheromones, secreted by the fish species. The mital cells present in the medial tracts of olfactory sense organs convey pheromonal responses to brain while lateral olfactory bundles are responsible for transducing feedback stimuli in fishes (Pal *et al.*, 2011).

The female fishes release pheromone along with ovarian fluid for pair formation, copulation, spawning, and courtship behavior of male. Androgen is essential to make male fish sensitive to the female pheromone. Maturation inducing hormones, estrogen, prostaglandin-F and steroid glucuronids are major components found in female pheromones. Odoriferous secretions of cutaneous anal gland, seminal vesicle, gill gland, sperm duct gland, urino-genital fluid of male fish serve as the source of pheromone. Testosteron steroid glucuronides and prostaglandin-F are major components of male pheromone.

7.2. REPRODUCTION IN SHELLFISHES

Decapods are egg-laying invertebrates. Most crustaceans are dioecius i.e. the sexes are separate and the sexes can be readily distinguished by the appearance of the anterior pleopods (swimmerets). The most anterior in female is abortive, but in male it, and the next one posterior to it, are modified to act as copulatory organ. But several groups like Remipedia and Cirripedia, are simultaneously hermaphrodites, while others are sequential hermaphrodites. In the latter, only one type of sex organ is active at a time and sex reversal occurs some time during the life span. So, a particular individual begins its reproductive activity either as a male (protandry) or a female (protogyny) and the sex is changed after one or more breeding activity. Some isopods and amphipods are protandric and several tanaidaceans are protogynous. A change of sex during the life time of an individual is a regular occurance in some Dendrobranchiates. Some individuals begin life as male but then transform into functional females after 13 months. Many branchiopods are parthenogenetic, the population mostly consists of females, which produce further females from unfertilized eggs. The males appear rarely, mainly under advese conditions, when fertilized resting eggs are produced.

The male prawns are bigger in size with narrower abdomen than females. The bases of thoracic legs are separated in females compared to males. The second pleopod in male prawns bears an additional process, the appendix masculine, in between the endopodite and the appendix internal. In males, the epimera of the abdominal segments are smaller than the females. The paired genital openings remain on the coxae of 5th pair of legs in males but in females these are on the coxae of 3rd pair of legs.

The blue claws mean the crab is a male and red tips on the claws mean it is a female. Males of crab and lobster species are normally larger than the females and posses much larger pincers. The males also have clasping organs used to hold the female during mating. The abdominal flap's apron, which is at the ventral side of cephalothorax, is slender and triangular in males but it is broad and semicircular in females (Fig-1.4c & d). The appendages are present only on first and second abdominal segments in males and modified for copulatory organ and sperm transfer process. There are four pairs of abdominal appendages in females from second to fifth segment, and are used to carry the eggs. The claws are comparatively larger in males than females.

7.2.1. Reproductive System

In most crustaceans, the gonads are elongated, paired organs, lying in the dorsal portion of the trunk. The oviducts and sperm ducts are paired, simple tubules that open either at the base of a pair of trunk appendages or on a sternite. The segments that bear the gonopores, however, vary from one group to another. Copulation is the general fact in crustaceans, the male usually having certain appendages modified for clasping the female.

The gonads are similar in position, shape, size and general disposition, in both the sexes of prawns. They are in the posterior region of the thorax dorsally above the hepatopancreas and below the pericardium. They extend anteriorly up to the renal sac and posteriorly up to the first abdominal segment. The cavities of the gonads represent true coelom whereas the gonoducts represent the coelomoducts.

7.2.1.1. Male reproductive organ

The soft, white paired but connected testes of decapods usually lie in the thorax above the hepatopancreas and may extend diverticula posteriorly into the anterior portion of the abdomen (Fig-7.5a). The testis is composed of six coiled lobes in dendrobranchiate shrimps (Fig-7.5a). All the lobes are connected at the inner margin loading to vas-deferens (Fig-7.5a). These lobules or seminiferous tubules are embedded in connective tissue. The cavity of each lobule has a single layer of germinal epithelium, the cells of which undergo spermatogenesis to produce spermatozoa (Fig-7.5b).

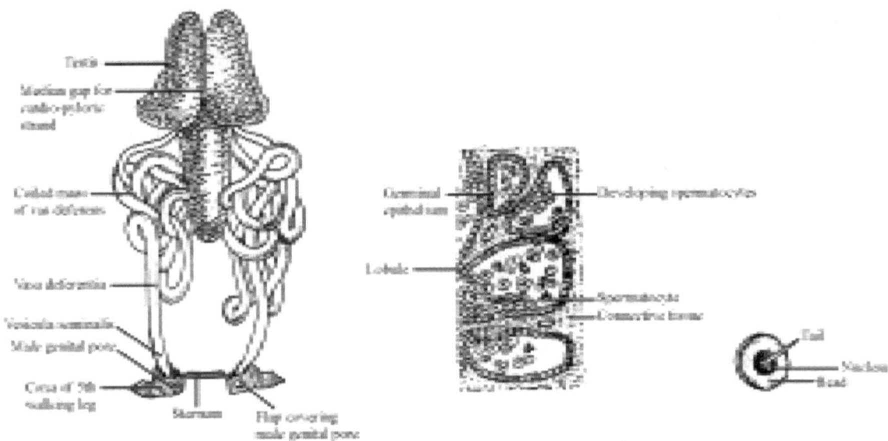

Fig. 7.5 : Prawn male reproductive system (A), histology of testis (b) and mature sperm (c).

The testes of caridian shrimps are simple tubes, connect to one another anteriorly. The anterior ends of the testes join to form a common lobe enclosing a gap for the passage of the cardio-pyloric strand, connecting the heart and pyloric stomach. The vas deferens or sperm duct of each side is much coiled at the origin and then runs vertically downwards between the abdominal flexor muscles on the side and thoracic wall on the outer side. The vas deferens has four regions such as - a short and narrow proximal region; a thick and larger medium portion; a long but narrow distal part and lastly a muscular ejaculatory portion known as terminal ampoule, which opens at the base of the coxa of fifth pereiopod or the articulating membrane between the coxa and sternum. The developing spermatids or spermatocytes are usually round to oval within the testes and mature to spermatozoa or spermatophore that are accumulated and stored in club-shaped seminal vesicle. The seminal vesicle is formed due to the swelling of vas deferens near the base of the fifth walking leg. Decapod sperm have a central spherical body with tail, and are star-shaped and non-motile.

The spermatozoa have three distinct regions - the cell body, cap and spike (Fig-7.5c). The cell body contains the typically uncondensed nucleus, which is not bounded by nuclear envelop. Mitochondria are also found. The cap region contains electron dense fibrils of varying size and contrioles in the central part of the spike. Acrosomal complex has been reported for some dendrobranchiate shrimps (Kleve *et al.*, 1980). The acrosomal vesicle is bilayered (inner and outer) in most brachyuran crabs and may be flanked by a prominent lamellar region. The acrosomal tubule may be supported by microfilaments or microtubules depending on the species. The spermatozoa of crayfish are different. The acrosomal vesicle is horse-shoe shaped, single layered and crystalline in nature. Acrosomal tubules are much reduced and microtubules are not visible. The cell membrane is much thicker than most other decapod spermatozoa.

The sperms may be transmitted in the form of spermatophores, in which case the more distal portion of the sperm duct is glandular and modified for spermatophore formation. The spermatozoa of tiger prawn are small in size. Each seminal vesicle open to the exterior by a male genital aperture on the inner side of the base of the coxa of the fifth walking leg. The first two pairs of pleopods in male decapods are usually modified to aid in sperm transfer. For example, in brachyurans the first pleopod is in the form of a cylinder into which fits the piston-like second pleopod. Sperm emitted from a paired penis in all brachyurans (single tubular in some hermit crab) are pumped through the conducting anterior pleopod.

In the mature male crab, the testis has two narrow convoluted arms. These are connected by a cross bridge posterior to the pyloric stomach. Posteriorly the testis connects on each side with much coiled vas deferens. The extent of coiling of vas deferens varies greatly with the developmental stage and differentiated into distinct regions. When fully developed it comprises four main divisions. The vas deferens runs latero-posteriorly to the opening at the tip of a papilla on the last thoracic coxa. This papilla lies permanently inserted into the groove of the first pleopod. A part of the second pleopod also fits into the groove and is believed to push the sperm to the female genital apperture. The mature spermatozoa have a central part with several radiating, finely pointed, cytoplasmic processes. The first region of the vas deferens secretes the thin capsules for the spermatophores and probably also at least a part of the material for the sperm plug. The spermatophores are stored in large numbers in the later thickers portions of vas deferens. The second region of vas deferens secretes further material for the sperm plug. Another constituent of the sperm plug may be added by tegumental glands in the first pleopod.

7.2.1.2. Female reproductive organ

The female reproductive organs consist of paired ovaries and oviducts, and a single thelycum. The former two are internal and the later one is external (Fig-7.6). In dendrobranchiates, the ovaries are located in the dorsal portion of the cephalothorax in the same relative position as the male testes i.e. lying dorsal to the hepatopancreas.

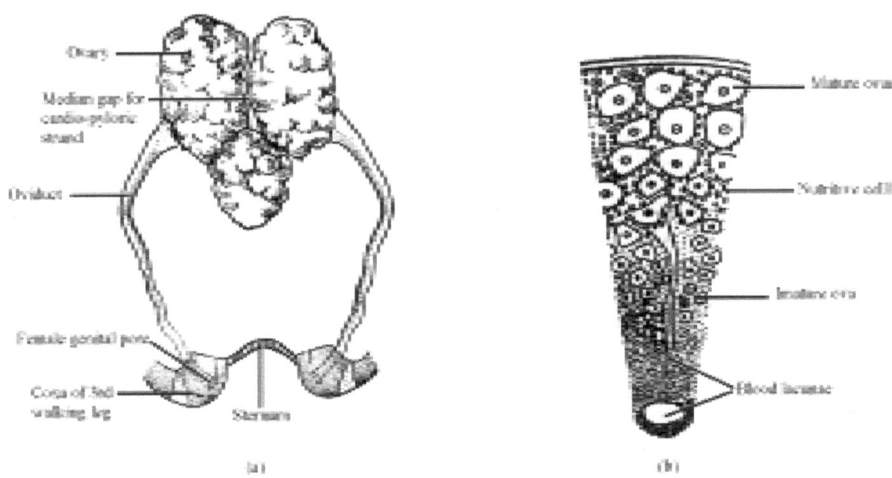

Fig. 7.6 : Prawn female reproductive system (a) and histology of ovary (b).

The ovaries are bilaterally symmetrical and partially fused leaving a gap in the middle for the passage of the cardio-pyloric strand. The size of the ovaries depends on the age and reproductive condition of the individual. In mature females the ovaries are greatly enlarged and extend into the abdominal somites. In the caphalothoracic region the organ bears a slender anterior lobe with five finger-like lateral projections. A pair of lobes, one from each ovary, extends into the abdomen. The anterior lobes remain close to the esophagus and cardiac region of the stomach. The lateral lobes are located in the large mass of hepatopancreas and ventral to the pericardiac chamber. The abdominal extensions lay dorso-lateral to the intestine and ventro-lateral to the dorsal abdominal artery. Each ovary is covered by a membranous capsule and is made of numerous radial rows of ova at various stages of development. The smaller immature ova lie towards the center while the mature large ova remain towards the surface of the ovary. The mature ova are nucleated with plenty of yolk material. The eggs are typically centro-lecithal. The short but wide and thin-walled oviducts originate at the tips of the sixth lateral lobes and descend to the exterior by female genital aperture hidden in the lobes of the coxopodes of the third pair of periopods. The thelycum is located between a pair of the fifth pleopods and consists of an anterior, and a pair of lateral plates.

In brachyuran crabs, the ovary has two lateral divisions, connected by a cross bar behind the pyloric stomach. These divisions run both forward and backward, above the hepatopancreas. The anterior portion on each side bends laterally over the cardiac stomach, and the fully developed ovary continues laterally behind the anterior margin of the cephalothorax to a point just posterior to the lateral spine. The posterior portion may extend back into the first abdominal somite when fully grown. Behind the cross bar, under the pericardial sinus, there is a pair of short oviduct, mostly concealed within the ovary. The duct on each side enters an oval shaped seminal receptacle near its medial antero-dorsal border. This receptacle extends downwards almost to the sternal floor of the thorax. The female genital apperture leads from postero-ventral border of the receptacle to the opening on the sixth thoracic sternum. The genital aperture and the ventral part of the receptacle have a cuticular lining, continuous with the exoskeleton. At immature stage, ovary is thin and white, and its arms may be very short. The receptacles are white and not larger than pyloric stomach. As the ovaries grow, they become orange in color and their arms reach full length. Copulation occurs just after final molt. Fertilization of the ova occurs at or shortly before egg-laying, either in the oviduct or in the seminal receptacle. In macurous forms, the ova pass from the ovary down the oviducts and exist via gonopore on the 3rd walking legs (pereiopods). In brachyurous forms, the short oviduct leads to a sack-like spermatheca within the musculature of the second walking lets.

7.2.1.3. Size at first maturity

According to Motoh (1981) sexual maturity of decapods is defined as the minimum size at which mature gaments are found inside the terminal ampullae of vas deferens in male and inside the thelycum of female. At 35 gm wet body weight shrimps attain maturity. The size of first maturity of crabs is between 84 to 120 mm carapace width.

7.2.1.4. Reproductive signals

In most decapod species, the males and females live separately from each other and pair-up as adults only to mate (Salmon, 1982). Since males and females are separated by vast spatial distances they rely on a combination of chemical, visual and acoustic signals to attract a mate (Salmon, 1982). Such signals are as follows:

1. **Chemical signals** - This type of attraction involves sex pheromones which are the most commonly used signal by aquatic decapods (Salmon, 1982). The use of pheromones is most evident in decapod species where female molting and mating are closely linked temporarily (Salmon, 1982).

2. **Visual signals** - This type of attraction occurs most commonly in close-range courtship amongst aquatic species or long-range courtship amongst land species of decapods (Salmon, 1982). Visual signals may involve a 'dance' or repetitive movement of a body part, such as "waving" (Salmon and Hyatt, 1983).
3. **Acoustic signals** - This type of attraction occurs predominantly in ghost and fiddler crabs (Salmon and Hyatt, 1983). These acoustic "calls" that occur only during the mating season are strictly produced by males to indirectly attract females (Salmon and Hyatt, 1983). Such signals in different species are as follows:
    a. Shrimps and prawns - Chemical attraction signals produced when the female molts to signify that she is sexually receptive and ready to mate.
    b. Spiny lobsters - Chemical attraction signals released during female molting to attract male. Fertilization occurs when a male first places a spermatophore on the female's sternum; her unfertilized eggs will pass over the area and become fertilized.
    c. Lobsters and crayfishes - Courting pattern varies greatly between the species. Courtship behaviors involve a combination of pheromones and molting depending on the species.
    d. Crabs - Marine species attract mates using chemical signals (e.g. sex pheromones). Courtship behavior lasts longer compared to other decapods species. Once paired together, the male will help the female and assist in her molting. During this process, the male will protect the female from predators and other males that may try to copulate with her.
    e. Mole crabs, hermit crabs and coconut crabs - Mole crab males experience neoteny and are physically smaller than females. This sexual size dimorphism allows the males and females to share the same niche therefore allowing easier access to mates during the breeding season.

7.2.1.5. Spawning and fertilization

Mating in most decapods occur shortly after female molting and the sexes are attracted to each other by pheromones or various other signals before or after molting, depending on the group. The spermatophores are simply deposited on the female sternal surface. In many decapods, however, this surface has become specialized, commonly as spermathecae (seminal receptacles), that receive spermatophores by way of spermathecal openings. In the higher groups of crabs, the oviduct connects directly with the spermathecae and uses their openings as gonopores. Lobsters and anomuran crabs transmit the spermatophores directly from the male gonopore to the female sternum. Some type of precopulatory courtship occurs. For example, the male hermit crab holds the female with one cheliped and strokes and pulls her with other. In some brachyuran crabs (*Cancer, Callinectes*), there is a premolt help by male to female in which the male carries the female till she molts and copulation occurs immediately afterwards. Male crabs take female into their burrows for mating and females remain there until their eggs hatch. Fertilization is internal in brachyurans. The courting of tropical species is restricted to daylight hours and, depending on the species, reaches a peak each month whenever the low tides occur with certain periods. Spermatozoa are stored within the spermatheca following copulation. The eggs are fertilized as they pass onto the abdominal pleopods for brooding (Warner, 1977).

Shrimps release their eggs directly into the seawater. In all other decapods, the eggs are typically attached to the pleopods with cementing materials that is associated with egg membrane (Felgenhauer and Abele, 1983). In most decapods, the eggs are probably fertilized at the moment of egg lying. In crayfish, the females remain on her back and curl the abdomen far forward, creating a chamber into which the eggs are driven by water current. In crabs, the tightly flexed abdomen is lowered to permit brooding and the orange colored egg mass is sometimes called a sponge. The hatching stage varies greatly. In shrimps the hatching stage is a naupliar or metanaupliar larva. In other decapods the eggs are carried on the pleopods of the female. In marine species hatching takes place at the protozoa and zoea stage.

7.2.2. Hormonal Regulation of Reproduction

An important neurosecretory system is located in the shellfish eyestalk i.e. X-organ-sinus gland complex. This complex produces neurohormones that regulate important physiological functions related to reproduction and other activities. The neurohormones affecting reproduction are - gonad inhibiting hormone (GIH) and molt inhibiting hormone. GIH is also known as vitellogenin inhibiting hormone as it inhibits vitellogenesis in females. GIH is also reported in male species. If the eyestalks are ablated, these inhibiting hormones are not produced and the oocyte maturation is quicker. As the eyestalks are removed, all the physiological activities, controlled by the other neurohormones of the complex such as movement of pigment in chromatophores, regulation of calcium during formation of cuticle, production of inhibiting hormone meant to inhibit the secretion of molting hormones of Y-organ affecting water regulation during ecdysis, inducing hyperglycaemia in blood to reduce stress, regulating lipid and protein metabolism in hepatopancreas, shall be stopped thereby adversely affecting normal physiology and growth of brood. The Y-organs lie in the maxillary segment of decapods and are the source of molting hormones or ecdysteroids, which promote molting and interact with molt inhibiting hormone from the X-organ. Thus, instead of removing both eyestalks, unilateral eyestalk ablation is generally practiced.

The most accepted model for crustacean reproduction is vitellogenesis (accumulation of vitellin or yolk protein) generally termed as ovarian maturation, is stimulated by gonad stimulating hormone (GSH) secreted by neuroendocrine and endocrine tissues, and inhibited by GIH of eyestalk. With the decline in GIH the species grows or moves into a suitable environment for spawning or sufficient nutrient reserves attained for vitellognesis.

It was reported that GIH is a peptide of 79 amino acids. It is expressed in brain, thoracic nerve cord in addition to the eyestalk. The inhibitory effect of GIH is due to the competition of vitellogenin and GIH molecule for the same receptor on the oocyte membrane (Jugan, 1985). GSH or gonadotropins has been found in brain and thoracic ganglion of several crustacean species (Gomez, 1965). Kleijn and Van Herp (1998) suggested that this hormone might be a crustacean hyperglycemic hormone. Methyl farnesoate (MF) (a crustacean hormone) is an intermediate compound produced during the juvenile hormone biosynthesis pathway. Juvenile hormone as such is not found in crustaceans (found in insects), but MF is isolated from mandibular organs of crustaceans. Mandibular organ actively synthesize MF during vetellogenesis and become less active during non-reproductive periods. The secretion of MF is tissue specific and circulated through blood, and circulating levels are positively correlated with the reproductive state of females. It has also been reported that mandibular organ is negatively regulated by an eyestalk neurohormone, mandibular organ inhibiting hormones (MOIH) and the inhibition is reversed by eyestalk ablation. The effect of MF on maturation is still not clear, although in many species a positive effect has been observed.

The brain and thoracic nerve centers produce hormones for development the sex organs. In addition, certain glands, attached to the male reproductive ducts, control the development of the male reproductive system. Their removal from a young male will cause it to develop into a female. The female ovary also acts as an endocrine organ; its endocrine secretions control the female reproductive system. The brood pouch of crustaceans also develops under the influence of ovarian secretions.

CHAPTER - 8

Sense Organs

The highly specialized organs that receive physical and chemical stimuli from the environment and various body parts, and transform them into stimuli are called sense organs or sensory receptors. These organs are closely associated with nervous system. The physical stimuli such as change of heat, light intensity and quality, acoustical stimuli and chemical stimuli like hear, taste, smell, touch, lateral line etc are received by the sense organs such as ear, eye, nose and others. These are more prominent sense organs. Less obvious ones in fishes are sensory crypts and papillae, ampullae of Lorenzini and Savi.

The crustaceans have various sense organs whose action are mediated through the nervous system. Crustacean antennae are important sites for the reception of environmental information, mainly chemical, and aesthetases, the olfactory hairs on the first pair, are usually well developed. The aesthetases help not only in locating food but also in recognizing other individuals and their sexual state. The frequent grooming of the antennae by many decapods prevents fouling of the receptor sites (Bauer, 1971).

8.1. SENSE ORGANS OF FISHES

Fishes can sense their environment in various ways. Besides the senses of sight, hearing, smell, taste and touch, they also possess sensory organs for detecting stimuli such as water particles, displacement and electrical current. Thus, the aspects of sense organs covered here are photoreception (vision), mechanoreception (lateral line for the detection of the objects in the water, hearing and orientation), chemoreception (smell and taste) and minor sense organs.

8.1.1. Photoreceptor (Vision)

Even though fish eye is the primary receptor site of the light, the pineal organ (endocrine organ) is also sensitive to light and appears to have importance in the control of circadian rhythmicity (Kavaliers, 1979). The fish's eye differs from those of land vertebrates in several ways. Most fish see to both right and left at the same time as the fish have no neck and cannot turn its head. Fishes living in brightly lighted shallow water will have relatively small but efficient eyes. But a few fish species are born blind. Such species are some catfishes who live in total darkness in waters of caves and the whale-fish, which lives in deep sea. Fish eye is modified for the vision both in air and water, and it is without any eyelid. But some species (*Galeorhinus*) have nictitating membrane like eyelid while some teleosts have fatty adipose tissue layer like eyelids.

8.1.1.1. Eye structure

8.1.1.1.1. Cornea

The cornea of typical teleost eye is of constant thickness (Fig-8.1).

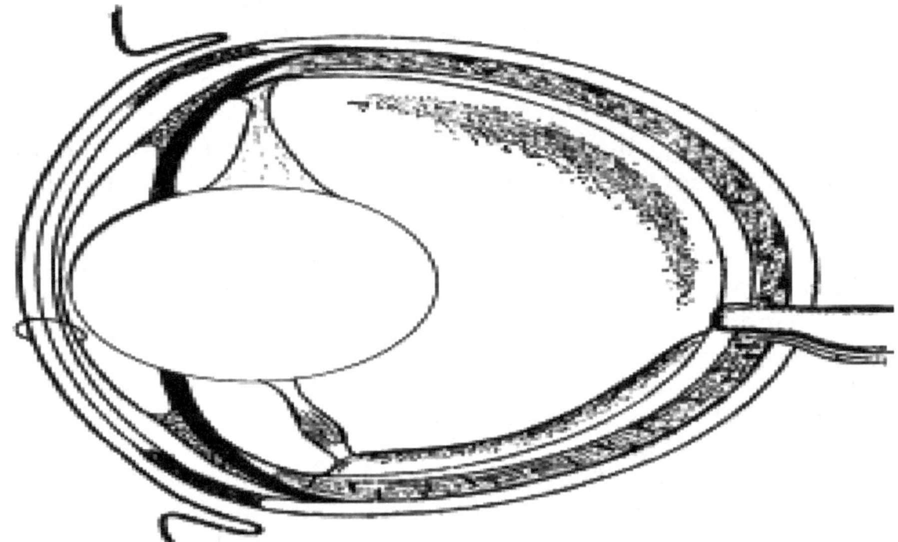

Fig. 8.1 : Diagrammatic representation of teleost eye.

It is the anterior most part of the eye and it cannot impose any optical alterations (convergence and divergence) on the incoming light. So, all the light focuses on the spherical lens. The cornea consists of corneal epithelium, stroma and endothelium. The cornea of most bony fishes is devoid of pigment, and so it is transparent. Other bony fishes such as *Perca* and *Amia* have green or yellow pigmented corneal epithelium. The cornea is double layered in *Solea*. In some teleosts a specialized inner autochthonous layer is present.

8.1.1.1.2. The sclerotic layer

This is the tough but highly vascularized layer that surrounds the eye ball. This layer is supported by fibrous tissue in elasmobranches. *Latimaria* has thick cartilage supporting sclerotic layer in addition to calcified scleral plates. In *Sturgeon*, this layer has a cartiligenous or bony support at its border. This layer is fibrous and flexible inside the orbit and around the optic nerve in teleosts.

8.1.1.1.3. The choroid or reflecting layer

This is present below the sclerotic layer. It is highly vascularized. In most bony fishes, the blood vessels are present in the shape of horse shoe in this layer, which is known as choroid gland. This gland secretes oxygen to meet the high oxygen demand of the tissue including retina. Generally, the retina does not possess well developed circulation. In eel, the choroid is supplied with less blood capillaries and choroid gland is absent.

8.1.1.1.4. The Iris

It projects over the anterior surface of the lens with its free edge forming the circular or oval pupil to control the amount of light that reaches the retina. The pupil is fixed as there is no sphincter and dialator muscle. In elasmobranches, the iris has muscle to adjust the pupil. The annular ligament is the circular connections between cornea and outer rim of the iris and the falciform ligament forms a ridge in the floor of the eye ball. The later ligament is not there in lungfish and some other fishes. Some guanine and melanin are present in the iris.

8.1.1.1.5. Lens

The small place between the cornea and the lens is filled with watery fluid called aqueous humour. The teleost eye lens protrudes out through the pupilar opening, and the eye bulges from the body surface enhancing the field of view. Light focused by the lens is projected in the appropriate pattern of light, shade, and often color, on the retina. The lens is firm, transparent and composed of non-collagenous protein. The lens is covered by a capsule and is filled by lens substance. In between the above two there is lens epithelium that plays an important role in metabolic processes of the lens. The lens substance consists of lens fibers, which are arranged in flat hexagonal prism manner. These fibers are modified epithelial cells. The bony fishes have spherical lens but in sharks and rays, the lens is horizontally compressed. In mud-skipper (*Periophthalmus*), the lens is flattened for aerial vision. The lens is pyriform in Anablepidae for aerial and aquatic vision. The lens is moved with muscular papillae present on ciliary body. Some elasmobranches accommodate by small changes in lens convexity. In trout, the lens has two focal lengths. One is for the light rays reflected from distant objects remaining lateral to the fish and focus on central retina. The other is for the light rays reflected from close objects and focus on posterior retina. So, such fishes have the capacity to focus distant and near objects simultaneously. The retractor lentis muscle contracts and brings the lens closer to the posterior retina that is concerned with the frontal field of fish vision. Fish lens has a high refractive index. The refractive index is from 1.53 at the center to 1.33 near the periphery. As the spherical lens is eccentrically located in the elliptical eye of teleosts, the distance between the lens and retina varies that determines relatively near-field vision in front of the fish and comparatively far-field vision to the sides. The near-sight vision coupled with binocular depth-perception capability allows the fish to capture their prey correctly.

The elasmobranches have unique ability to dilate and constrict the pupil of the eye, comprising another mechanism to regulate incoming light levels, although some deep sea sharks have fixed pupil (Kuchnow, 1971). The constricted, light-adapted pupil takes various shapes in different fishes like pinhole (*Carcharhinus*), vertical slit (*Negaprion*), horizontal slit (*Spyrna*) or oblique slit (*Ginglystoma*). Sawfish (*Pristis*) and ray (*Raja*) have an operculare or opaque projection, which descends vertically to cover part of the pupillary opening.

Adjustments in focus for both near and far vision are done by the movement of the lens by the muscle within the eye without changing its shape. The specific muscles used for eye movement differ in lampreys, elasmobranches and teleosts. The retractor muscle pulls the teleost eye lens inward while in elasmobranch the protractor muscle pulls the lens outwards. The four-eyed *Anableps* displays an aspherical lens and two-part retina as structural adaptations associated with its life at the air-water interface. The lens thickness differences allow for the density differences of the two media.

8.1.1.1.6. Retina

The retina is double layered. The outer layer is heavily pigmented and is known as granular (nuclear) layer. The inner layer is less pigmented. The retina has rod and cone visual nerve cells for discrimination of light. The rod and cones are densely packed in retina. Most fishes have duplex retina i.e. one containing both the rods and cones and the other is only with rods. The later is found in sharks, rays and deep-sea teleosts. In many larval fishes there is a pure cone retina and rods develop at later stage.

The incoming light must penetrate a clear layer of nerve cells and fibers to reach the photochemically active tips of the rods and cones. The nerve cells are horizontal cells, bipolar cells and amacrine cells. The horizontal cells are present in the peripheral region of the inner nuclear layer. These cells give rise to some processes that extend horizontally near the outer nuclear layer and acts as communicating lines between visual cells. The bipolar cells are present at the innermost area of the retina. These are the largest neurons of retina. The axons from these cells assemble and form optic nerve. Amacrine cells are found between granular and inner plexiform layers and act as horizontal lines for passage of visual stimuli. Gruber *et al.* (1975) measured the rod:cone cell ratio, which is approximately 10:1, for four species of sharks and about 6:1 in these active mackerel species.

Both rods and cones share a common plan i.e. the outer segment containing visual pigments, an ellipsoid packed with mitochondria, an extensible myoid or foot-piece and a nuclear region. These visual cells are held in position by an external limiting membrane. The rods exceed in number than the cones in adult fish eye. There are specialized areas in retina of many fishes. The area temporalis consists of a patch of closely packed cones with negligible or no rods, to receive light along the main axis of feeding. In pelagic horse mackerel and herring, which look upwards and forwards, the area is postero-ventral on the retina. In horizontally feeding fish (eg. *Istiophorus*), it is posterior on the retina, and in bottom feeders (eg. *Sparus*), the area is postero-dorsal on the retina. The mud-skipper (*Periophthalmus*) has a horizontal band of cones. A depression with higher number of cones is present in *Hippocampus* and *Syngnathus*. However, these visual cells are not clearly distinguishable morphologically in lampray and hagfish has degenerated retina.

8.1.1.1.7. Visual pigments

These consist of opsin protein linked to an aldehyde of vitamin A_1 or A_2. The A_1 pigments are called as rhodopsin and A_2 are porphyropsin. Each pigment have a characteristic light absorption efficiency. As these pigments are photosensitive, they are most sensitive to light of wavelength near the absorption maxima (λ). As they break down chemically, the pigments bleach and only regenerate to their original chemical state in the dark, a process that may take many hours to complete. These pigments are present in the outer part of both rods and cones. After bleaching some of the break down products migrate to the epithelium.

8.1.1.2. Functioning of eye

The vision is a photochemical process that involves the reactions in the light sensitivity pigments of the rods and cones. These chemical changes become transformed into the electrical impulses that can be noted in the retina and these are ultimately converted into signals, which is taken up by the optic nerve and go to brain as stimuli. As the retina is inverted, the visual cells are at the periphery of the eye and away from the lens. The epithelium of retina contains photosensitive melanin and borders the choroid coat. In bright light melanin spreads and shades the sensitive pathway. But in dim light it aggregates near the choroid layer, so that the photosensitive cells are fully exposed.

The rod myoids are shorter than cones, so that the rod outer segments are near to the external limiting membrane and the cones are not impending this penetration of light to the rods. During light adaptation the cone myoids shorten and the rod myoids lengthen taking their outer segments away from the cones and into the advancing masking pigment that protects them from the bright light. The teleosts have switch over adaptation from dim to bright light. This light and dark adaptation is accompanied by retinomotor or photomechanical movements of the rods and cones, and melanin masking pigment, akin to the occlusion of the tapetum. Light passing through the retina is reflected back by a reflecting tapetum present at the back of the eye. The tapetum is a specular reflector consisting of layers of reflecting cells in the choroid layer packed with thin platelets of guanine. The platelets are arranged at angles to reflect light back into the retina along the long axis of the visual cells. The reflectors of this type are adjusted to reflect certain wavebands by appropriate thickness and spacing of the guanine. The reflectors of deep-sea sharks reflect best at wavelengths that penetrate into deep water. In larval fishes, the retinomotor movements do not take place during the transfer from dark to bright light. The tubular eyes of teleosts point upwards and forwards with their optical axis more or less parallel. This type of eyes have large lens and have good binocular vision in one direction. The problem of this type of eye is that the peripheral parts of the retina cannot be brought into focus because they are very close to the lens. They are probably only useful for unfocused movement detection.

Water absorbs different colors of light selectively, some wavelengths being transmitted more rapidly than others. The wavelength (λ) which is best transmitted is the max 470-480 nm in deep sea, 500-530 nm near the coast and 550-560 nm or longer in most freshwaters. This change from blue to yellow is caused by the presence of yellow pigments (mainly the break down products of chlorophyll and humic acids in freshwaters) that has important implications in the performance of fish eye in different colored lights. The light rays pass through the associated nerve cell layers of the retina (the bipolar and ganglion cell layers) before reaching the visual cells. These layers are very transparent to light in order not to interfere with vision. The reflecting (choroid) layer passes back the incident light again through the visual cells. The refractive index of the cornea and occular fluids is similar to that of water. Refraction takes place entirely in the lens that is usually spherical with a short focal length. In teleosts, it is about 2.55 x radius of the lens, or Matthiessen's ratio. The refractive index varies across the diameter of the lens so that the rays follow a curved path. The retina is not always concentric with the lens and it has been reported that the resting eye is short-sighted anteriorly and far-sighted laterally. As the lens is moved posteriorly the lateral view changes very little but the fish would become able to focus distant objects anteriorly.

8.1.2. Mechanoreceptors (Acoustico-lateralis System)

This system helps the fishes to sense sounds, vibrations, gravity, and other displacements of water. It has two main components - the lateral line/neuromast system and inner ear. Fish inner ear functions to orient or balance the fish in three dimensional space giving it a feeling of the direction in which gravity is acting even when suspended in lightless, pelagic habitats. This system can respond to linear and angular accelerations of the fish's own body.

8.1.2.1. Lateral line system

This provides the fishes a ferntastsinn or distant touch sense (Dijkgraaf, 1962). By means of mechanoreceptors, similar to those in auditory and equilibrium systems, water movements around the fish can be detected. The presence of the object can be made out by fish from the reflected waves by echolocation. It involves sensory lines distributed on the head and body, pit organs and ampullae of Lorenzini.

The lateral line organs develop as thickening on both sides of the head near the ear. These thickenings then extend in definite lines of the body to form series of organs in different areas of the body. At points on these lines the sensory areas develop by the differentiation into two types of cells, the sensory and supporting cells. The sensory cells gather at different points to form a sense hillock or neuromast. The neuromasts are the receptors. The lateral line organs and their canals are variously distributed in fishes. A supra orbital line runs forward from the region of ear and above the eye to the tip of the snout. This line is supplied by the superfacial opthalmic branch of the 7th cranial nerve. An intraorbital also runs forward below the eye and is innervated by the buccalis nerve. A hyomandibular line passes along the lower jaw and operculum and is innervated by the mandibularis externus. The lateral line proper runs back on either side to the tip of caudal fin. The cranial nerve VII, IX and X innervates the lateral line at different points. These nerves have close association with auditory centers provided with cranial nerve VII. A supra temporal commissural line connects the canal system of both sides of the body.

There are two types of neuromasts. These are located in pits in the epidermis present primarily on the head. On body surface, the neuromasts are free, in lateral line system these are placed in pits or canals. These are in large fields (cristae) in the ampullary organs of the semicircular canals and in the maculae of the sacculus, utriculus and lagena of the inner ear. The neuromasts are usually interconnected by closed canals and are also connected with the exterior by other shorter canals which arise from the external pores present on the skin or scales. In head region the canals themselves find connections with the exterior by passing through some cranial bones. In general, active fishes have a greater percentage of canal neuromasts than free ones. Many larval fishes hatch only with free neuromasts and lateal line develops afterwards.

Each neuromast consists of individual hair cells with an attached cupula (Fig-8.2). The external bundle of hairs projects into a gelatinous cupula. The cupulae of the canal neuromasts are sensitive to movements of the watery endolymph fluid through canal. There is one long hair, known as kinocilium that has the typical 9+2 internal array of tubules found in all cilia. On one side of the kinocilium, there are a number of shorter microvilli or stereovillae (or stereocilia) which do not have any internal tubules. Water movements bend the projecting cupula, which stimulates the hair cell by bending the attached cilia (sense hairs).

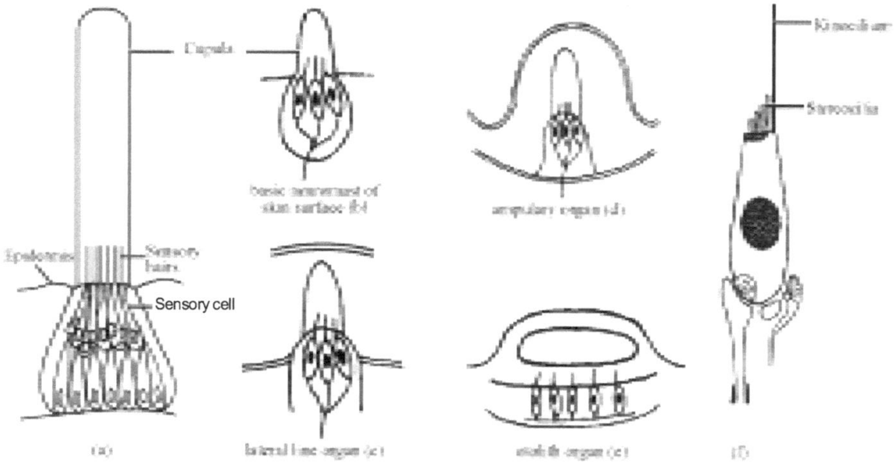

Fig. 8.2 : Structure of superficial neuromast of bony fish (a), basic neuromast of skin surface (b), modifications of neuromast where it is incorporated into lateral line (c-e) and single hair cell structure (f).

The hair cells not only synapse with afferent nerve fibers carrying information to the brain. They also receive inhibitory efferent fibers whose action is to switch off the cells. Like the auditory and equilibrium hair cells, the neuromasts continually send neural impulses to the brain. The neural impulse frequency is increased as the cupulae are flexed in one direction and reverse in other direction. The displacement of the cupula causes the stereocilia to bend towards the kinocilium, and then the hair cells become depolarized, causing excitation. But the cells are hyperpolarized with an inhibitory effect when the stereocilia are bent away from the kinocilium. The hair (sensory) cells are grouped into neuromast organs. Usually the hair cells in some areas of a neuromast organ are all polarized along the same axis.

The pattern of impulses from the free or canal neuromast imparts a direction to the disturbance. The head lateral line organs are supplied by sensory fibers of the lateralis anterior root of cranial nerve VII (facial). The other organs of the system are innervated by the lateralis posterior root of the vagus. The fibers of both roots join with labyrinth nerve (VIII) in the acoustic tubercle of medulla oblongata.

A sound source produces two types of stimulus - a back and forward motion of the particle in the medium (particle displacement) and a sinusoidal change in pressure (sound pressure). Their amplitudes drop at different rates depending on the distance from the source. Particle displacement is more important close to the source and sound pressure at a distance from the source. The area close to the source (particle displacement) is called near-field and the more distance region (sound pressure) is the far-field. To detect far-field effects, the sense organs are to respond to sound pressure.

Lateral line system is very much developed and used in various ways in fishes. Fishes of streams have extensive canal neuromast system, which is still more in active fishes, but those who are living in still-water has no canals (Alexander, 1967). The canal offers some protection from the continuous stimulation of water moving past the laterally located neuromasts. Thus, the canal-based receptors can still function to detect weak local water displacements during rapid swimming (Dijkgroaf, 1962).

8.1.2.2. Inner ear

Inner ears are present on both sides of the head and connected to each other by transverse canal. These ears were described by Weber in 1820. It is the main organ for hearing and balance. Some species (Ostariophysi) have their air-bladder connected to ear by Weberian ossicles. The inner ear of fish consists of a membranous sac enclosed in a chamber on either side of the posterior part of the skull. The sac can be divided into a pars superior or upper chamber consisting of three semicircular canals, one vertical anterior and two posterior horizontal with their ampullae and a sac-like utriculus, and pars inferior or lower chamber having two vesicles, the sacculus and lagena (Fig-8.3). There is a macular neglecta on the outer wall of the constriction between utriculus and sacculus. Each semicircular canal enlarges at one of its anterior ends to form ampulla and utriculus. In sharks, the anterior vertical canal joins with horizontal canals to form crus. The three semicircular canals open into the utriculus and are important in maintaining the balance. Lagena is a small out-growth of sacculus.

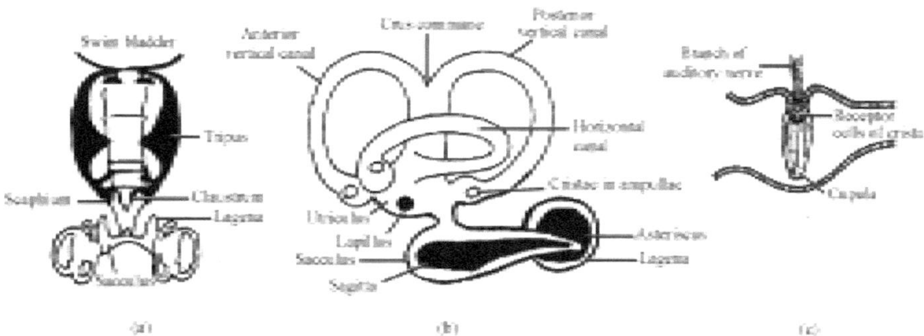

Fig. 8.3 : Structure of ear ossicle (a), membranous labyrinth (b) and crista (c).

The inner walls of the utriculus, sacculus and lagena are provided with groups of highly sensitive cells. Cavities of these three chambers are filled with endolymph and solid concretions, the otoliths. The perilymph is present in the space surrounding the membranous labyrinth between it and the auditory capsule. The ampullae contain patches of receptor tissue, the cristae staticae with sensory cells. The otoliths of utriculus, sacculus and lagena are known as lapillus, sagita and astericus, respectively. In bony fishes, these are calcified structures that are composed of calcium carbonate, keratine and mucopolysaccharides, and are secreted by ectoderm. The lapillus remains horizontally on the hair of the utriculus sensory cells and that of the other two remain vertically over sensory hair. To maintain body equilibrium they work together with the lower part of retina. So, eye receiving light from above and utriculi receiving gravity from below push and pull keeping the fish in an upright position. The endolymphatic duct is short and closed tube in teleosts. But it is absent in *Salmo* and *Lampanyctus* (Jollie, 1968). In elasmobranches, an endolymphatic duct originates from the inner ear and extends to outside. The sand particles enter into the ear through this duct and reach the gelatinous cupula of the hair cell. These particles act as otolith.

Pars superior and posterior are innervated by the anterior and posterior branches of the auditory nerve (VIII), respectively. The vesticular branch of the auditory nerve is distributed in a fan-like structure on the ventral surface of the utriculus and innervates it and ampullae of the semicircular canals. The saccular branch runs back and supplies the lagena, sacculus and ductus lymphaticus. The sensory cells are covered by tactorial membrane that transmits stimuli to the sensory hair by vibrations (Hibiya, 1982). This part is known as organ of corti.

The ear ossicles perform two major functions i.e. the sound detection and maintenance of balance. They help to orient the animal providing the fish to feel the direction in which gravity is acting when it remains in pelagic habitat. The semicircular canals and the utriculus are mainly responsible for the maintenance of equilibrium, while the sacculus and lagena are for hearing. Sound travels 4 to 8 times faster under water than air as it has greater density. Water is a much efficient conductor of sound pressure waves.

The cupula extends into the canal path, partially blocking the flow of endolymph. The angular accelerations, either from swimming or turning movements of the fish or from water currents moving the fish are detected as the endolymph lags behind the movements of the labyrinth. Proper cupulae are bent by the pressure of the endolymph and stimulate the hair cells. Thus, the sensory hair cell system as a mechanoreceptor parallels that of the hearing and lateral line system. Changes in the pattern of the continuous lines of neural impulses from the hair cells to the balance and equilibrium center in the medulla triggers the appropriate motor responses by the fish like eye movements for stable vision and fin movements to resume body equilibrium.

Sound waves, especially those of low frequencies, travel readily through water and impringe directly upon the bones and fluids of the head and body, to be transmitted to the hearing organs. Fishes quickly respond to sound. For example a trout conditioned to escape by the approach of fishes, will take flight upon perceiving footsteps on a steam bank even if it can not see the fisherman. The range of sound frequency heard by fishes is very much restricted. Many fishes communicate with other by producing sounds in their swim bladder, by rasping their teeth and in other ways.

8.1.3. Electroreceptors and Electric Organs

These receptors of fishes are able to pick up the small steady d.c. fields generated by living prey and the reception of minute electrical currents in the water. Electroreceptors of a different type, responding to high frequencies are found in several teleosts and some fishes have electric organs producing electric fields used in prey detection and social signaling. Specialization for electroreception is widespread among fishes, except some teleosts represented by relatively few groups (Bullock *et al.*, 1983). The electroreceptors are of two types - ampullary and phasic.

8.1.3.1. Ampullary (tonic) receptors

These consists of ampullae of Lorenzini and Savi with canals whose wall contains clusters of sensory cells. These are found in the head of elasmobranches and over the upper and lower surfaces of the wings of rays. These are closely linked to lateral line in nerve supply. Both are similar in structure. The ampullae has a low-resistance electrical conductive jelly-filled canal that opens to the external water medium through a pore in the skin and ending in a swollen bulb (Fig-8.4). Freshwater catfishes have very short (300μm in length) canals in contrast to marine catfish (*Plotosus*) with longer (5-160mm in length) canals. This canal length is related to the electrical conductivity differences between the environments compared with those of the skin and body tissues. The relatively low resistivity of the skin and the greater resistivity of internal tissues compared with the sea-water environment make the marine fishes quite transparent to external voltage gradients. But freshwater fishes have relatively good conducting tissues in comparison to their environment and their skin has high electrical resistance, causing large voltage drops compared with the other tissues. The canal may have some diverticula lines with sensory epithelium.

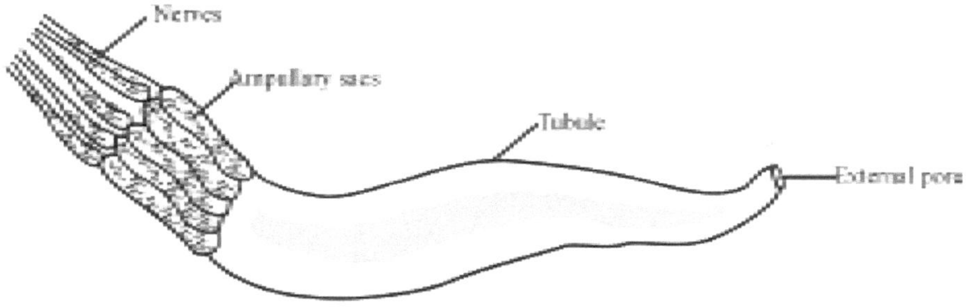

Fig. 8.4 : Diagrammatic presentation of Ampullae of Lorenzini.

Two types of sensory cells, pyramid and flask cells, are present the ampullary wall. The apex of both sensory cells, bears microvilli and/or a single kinocilium protruding into the luman of ampulla. The canal wall has high resistance which is 30-100 times more than that of the nerve myelin sheath. The ampullary organs are tonic receptors giving a long lasting response to very low frequency (or d.c.) stimuli. Although they are modified neuromasts, the sensory cells appear to have no efferent innervations. The sensory cells are supplied with facial nerve fibers.

The ampullae help the fish to compare electric potentials over a considerable area improving their ability to detect the direction of the source of an electric stimulus.

8.1.3.2. Tuberous (Phasic) Receptors

This type of electroreceptors are tuberous organs consisting of epidermal capsules with no connection to external medium due to lack of a canal. These receptors adapt quickly and are better suitable for the reception of high frequency from electric organs. The apex of sensory cells has no kinocilium and the microvilli are closely packed. These receptors are phasic as they give brief responses to high frequency electrical stimuli. These receptors have low resistance pathway to the skin surface due to the absence of canal.

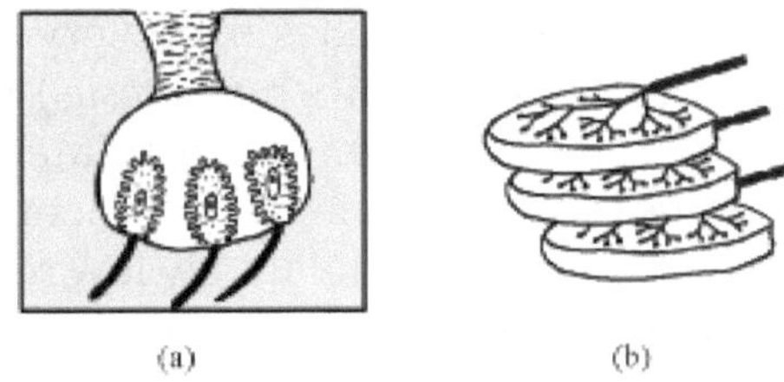

Fig. 8.5 : Electric organ of fish - tuberous mormyrid receptor with large surface area of receptor cells within cavity (a) and stacked modified muscle fibre that are innervated into the electric organ of electric fish (b).

8.1.3.3. Electric organs

Kalmijn (1971) reported that predaceous marine elasmobranches locate their prey by electroreception alone. Mostly fish electric organs are modified striated muscle fibers and made-up off layers of flattened cells innervated on one side (Fig-8.5). Marine stargazers (*Astroscopus*), has its electric organ developed by the modification of its eye muscles. In catfish *Malapterus,* the electric organ is situated between the skin and muscles. In *Torpedo,* two electric organs are present. But in freshwater sternarchids, these organs are derived from modified nerve fiber. The arrangement of flattened cells adds together the small electric potentials generated from membrane depolarizations to give rise to higher external potentials. Freshwater eel (*Electrophorus*) are able to produce 500 v, which acts as an effective defense against predators and is sufficient to kill the prey. In many species these organs function by electrolocation mechanism similar to echolocation. Weakly electric fish can recognize objects in the medium by distortions they produce in the electric field generated by their electric organs.

8.1.3.4. Functions of electroreceptors

These receptors are used by the fish for self defense, and to locate and kill the prey. Any electrical conductor moving through a magnetic field induces an electrical field through the conductor. Thus fish having electroreceptors dorsally and ventrally, and swimming across the earth's north - south magnetic field should be able to detect the induced electrical currents. These induced currents are of sufficient strength for detection by shark's swimming as slowly as 2cm/s, proving the feasibility of an electromagnetic compass sense (Kalmijn, 1978). It is further reported that the east bound shark when turns north or south, the potentials vanish. Turns towards west produces potentials of the opposite polarity. Walker *et al.*(1984) found magnetite crystals in a dermethmoid bone sinus of the yellowfin tuna (*Thunnus albacares*), a fish known to behaviorally respond to earth-strength magnetic fields. It is possible to pick-up and record signals produced under different conditions. Some species (mormyrids and gymnotids) increase the frequency of their electric pulses upon stimulation and in some situations the discharges stop totally. When the fish is present in high density, there is a problem of interference between the electric discharges of the animals close to one another.

Marine ray (*Torpedo*) is weakly electric with caudal electric organs and does not use these organs for electrolocation and perhaps use the electric discharges for intra-specific recognition. The electric discharges of stargazers are insufficient for electrolocation. During feeding the electric organs produce a brust of high frequency pulses of 150 - 300 ms and the duration of the burst is for a period till the mouth is open.

8.1.4. Chemoreceptors (Smell and Taste)

In fish, the smell (olfaction) and taste (gustation) are of similar sensitivities, to contact stimuli of chemicals dissolved in water but the organs for these reception are distinguished by the location of the sensory receptors as well as processing centers in the brain. It is generally assumed that olfaction is distance reception and gustation is contact reception. In fish the taste buds are not only found in mouth but also on the gills, head surface, fin and barbels. In some fishes these are found on whole body surface. Carps have large number of taste buds in their pharynx.

8.1.4.1. Taste buds (Gustation)

Many catfishes, which do not feed by sight, have large number of taste buds (Fig-8.6) on the lips and barbels. Fish taste buds have high sensitivity to a single amino acid or a combination of amino acids. A taste bud is formed of neuroepithelial sensory and subtentacular supporting cells. The sensory cells have small hair-like extensions at their free ends and the basal ends are embraced by the nerve fibers from cranial nerve VII, IX and X. The cutaneous taste buds are innervated by branches of facial (cranial VII) nerve. The sensory signals from the internal taste buds (eg. those in the pharyngeal cavity and palate) are transmitted to the glossopharyngeal (IX) and vagus (X) nerves. The cutaneous taste buds may trigger the pick-up behavior but the food may be rejected after tasting in mouth. Lateral line and/ or electroreceptors may also play a role in this pack-up. In species having a palatal organ, the total number of taste buds is considerably increased, and the terminal centers in the visceral sensory column of the medulla are correspondingly enlarged as a vagal lobe (Kapoor *et al.,* 1975). The sharks and rays show no sensory elaboration for taste functions (Kapoor *et al.,*1975). Elasmobranches rely mainly on olfaction, vision and electroreception to locate their food.

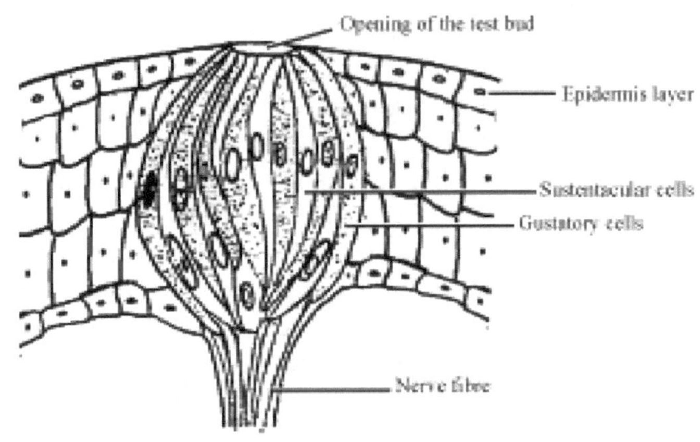

Fig. 8.6 : A test bud of teleost fish.

8.1.4.2. Smell (Olfaction)

Fishes are able to detect the odour with the help of a pair of oval shaped olfactory rosettes in a chamber and connected to the olfactory lobe of the brain by means of olfactory nerves. There is a wide variation in the structures of the olfactory organs of fishes. The olfactory chamber is present on either side of the head and is placed in a depression in the ethmoid bone between it and the palatine and lacrymal bone. Lamprey and hagfish have a single chamber and single nostril. In hagfish, there is a nasopharyngeal duct opening into pharynx but in lamprey this duct ends in a nasopharyngeal pouch. In hagfish, the nostril opens into the oral cavity.

The olfactory receptors are usually located in ciliated olfactory pits, which have incurrent and excurrent nostrils or channels divided by a flap of skin. Due to the movement of cilia, by the muscular movement of the branchial pump, by swimming or by a combination of these, water flows into the olfactory pit to create a pressure difference between incurrent and excurrent nostrils. The olfactory sensory epithelium consists of four types of receptor cells such as olfactory (receptor), supporting (subtentacular), basal and mucous (secretory) cells. Sensory cells are uniformly distributed on both sides (Khanna and Singh, 1965). Two types each of receptor and supporting cells have been distinguished in *Barilius bendelisis* (Singh, 1982). The columnar supporting cells may or may not be ciliated. The receptor cells are distinguished as primary and secondary neurons (Hora, 1986). The dendrites of each primary receptor extend up to the free epithelial surface where it enlarges into an olfactory vesicle that supports the sensory hair. The axonal ends of these receptor cells synapse with dendrites of secondary neurons. The secondary neurons generally present in the deeper part of the epithelium with their axonal ends extends upto the basement membrane. The basal cells are either cuboidal with fine cytoplasmic processes or round with oval nucleus. A few sensory goblet type cells are present in the epithelium. The basal cells give rise to other cell types and replace the degenerating cells (Pandey and Misra, 1984).

Intense alkaline phosphatase and less acid protease activity has been observed in the receptor cell dendrites (Pandey and Misra, 1984). The activity of these enzymes is low in the supporting basal and secretory cells. Lipid is widely found in olfactory epithelium. The mucous secreted by the mucous cells contains mucin protein and acid mucopolysaccharides and forms an uniform layer around the olfactory epithelium to provide protection to sensory hair against osmotic effect of water.

The olfactory stimuli (via lateral or medial divisions of the olfactory bulb) are communicated to the olfactory lobe of the brain through the first cranial nerve. According to Khanna and Singh (1965), the carnivorus and predatory fishes possess better olfactory sense, generally having an elongated type of rosette and a large number of sensory cells. The herbivorous fishes, in contrast, possess a weaker olfactory sense. Olfactory cues are important for the migratory fish like salmon to locate their native stream. Thus salmon are thought to be imprinted with the odors as presmolts and smolts as they migrate down the rivers and streams to a larger water body (Haster and Scholz, 1983)

8.1.5. Photophores and Bioluminescence

Many deep-sea fishes possess light producing organs or photophores of different types to produce characteristic luminescence (bioluminescence). Some species use lights as lures. Fishing rods with luminous tips are found in several families, whilst others have light organs in the mouth. The flashlight fish (*Photoblepharon*) uses them as headlights to illuminate their prey. Most fishes use their photophores to send signals to other members of the same species.These organs are probably specialized gland cells of the epidermis. Mostly they occur along the lateral and ventral sides of the body and head. They may be arranged in one or two rows extending on the sides from the head to tail or they may be present on certain patches of the body. Large photophores also occur on the elongated first fin rays of pectoral and dorsal fins.

Luminescent organs or photophores are of two main types: those which are self-luminous to produce light by special photocytes and in others the light is produced by symbiotic bacteria present in special sacs. Organs of the latter type are found in a large number of species. Bacterial light organs are either linked to the gut or open to the exterior. There are diverticula of the gut and are surrounded by connective tissue reflecting layer. They are specific to each host fish. They glow continuously and the fish can only mask them or rotate them into a black-lined pocket. Some other bacterial light organs are very dim and are like the peri-anal organs of *Chlorophthalmus*. These may be used as clue for schooling. Some fish use these organs for camouflage.

Structurally, these photophores consist of a large number of glandular tubules that secrete luminous bacteria. The self-luminous simple photophores consist of a series of radially arranged glandular tubules that are innervated by cranial and spinal nerves, and light is generated intrinsically. In more complex forms, a thick reflecting spicule is present in the cup and a lens like structure is also developed on the mouth of the cup. Besides the simple ones, there are complex sub-orbital cup-like photophore organs made up off many concentric layers. Externally, there is a layer of black pigment and numerous glandular tubules are present in the cup. This organ is supplied by a branch of cranial nerve (V). The luminescent organs contain luciferin and enzyme luciferase. Luciferin is an indole derivative containing tryptamine, arginine and isoleucine. Luciferase oxidizes luciferin to oxyluciferin that emits the light.

8.2. SENSE ORGANS OF SHELLFISH

In shellfishes, the legs and antennae are important sites for the reception of environmental information and aesthetases. The chemosensory hairs on the first pair of antennae are commonly well developed besides that of the compound eye and staloliths. The frequent grooming of the antennae by many decapods prevents fouling of the receptor sites. Grooming is a common behavior found in many crustaceans as a form of maintenance to rid their bodies from epizoic growth and particulate build-up. In decapods, this behavior is extremely crucial and helps to prevent the growth of foreign organisms on their body surfaces, loss of olfaction hairs, clogging of gills and embryo mortality. Grooming within the decapod crustaceans involves the use of chelipeds and the setal brushes at the tips of their legs. The most frequently groomed areas of the body include the sensory and respiratory structures. Some decapods use mechanisms other than grooming to maintain their body surface clean. These alternative methods includes :

a. Burrowing into deep sediments.

b. Wedging between narrow surfaces.

c. Changing temporal lifestyle from diurnal to nocturnal.

d. Molting.

8.2.1. Eye and Vision

Light detection and vision are important senses in many crustaceans. Some small species and many larval forms have only a simple nauplier eye, consisting of a few oceli grouped together medially on the head. Although there are some blind decapods, particularly among deep-sea forms and cave dwelling crayfish, paired compound eyes are usually highly developed. The eye stalk or occular peduncle may be sessile or highly mobile as in Malacostracans. The peduncle is composed of two or rarely three segments. In some crabs, one of the peduncular segments is greatly lengthened, placing the eye at the ends of extreamly long stalk to which a series of light-sensing organs are attached. Nearly all surface dwelling shellfishes have pigmented eyes. The multiple faceted compound eye of crabs and shrimps can be lowered into sockets on the carapace for protection. The snapping shrimps are peculiar in that the anterior margin of the carapace has grown down and completely encloses the eye. However, the covering is probably thin enough to permit light perception. In brachyurans, there may be transverse orbits in which the eyes rest.

The decapod's eyes are surmounted by a black colored hemisphere lying beside and slightly below the rostrum. It can be swivelled to some extent by muscles, attached internally to the cuticle of the stalk. By means of these muscles, the eye can be directed towards the object in view. The black hemisphere is composed of large number of optically sensitive units or facets known as ommatidia (Fig-8.7). Each ommatidium is shaped like an elongated cone with the apex directed inwardly so that they fit to each other to form a fan-like structure. Due to the presence of multiple ommatidia, the surface of the eye hemisphere is mosaic. The ommatidium is multicellular structure. In each ommatidium four flattened epidermal cells are present at the outer end, which secrete externally a transparent cuticle forming a cuticular lens (Fig-8.7c). Below the epidermal cells there are four elongated cells known as vitrellae. These cells combine to secrete and to surround a conical refractive lens or crystalline cone. Surrounding the lower ends of the vitrellae, there are six retinular cells forming the retinula (small retina) and enclosing a spindle-shaped transversely striped part called the rhabdome. The rhabdom contains the clentrites of the eccentric cell and may contribute some microvilli. These retinular cells are photoreceptors and nerve fibers of the optic ganglion nerve cells are innervated into these cells. Surrounding the vitrellae and retinula are the outer and inner iris cells having black melanin pigment, which optically isolates each ommatidium from the next one. The corneal surface is usually terminal, but this is not always the case. In some crabs (*Ocypods*), the peduncle extends beyond the corneal surface, which is wrapped around the stalk.

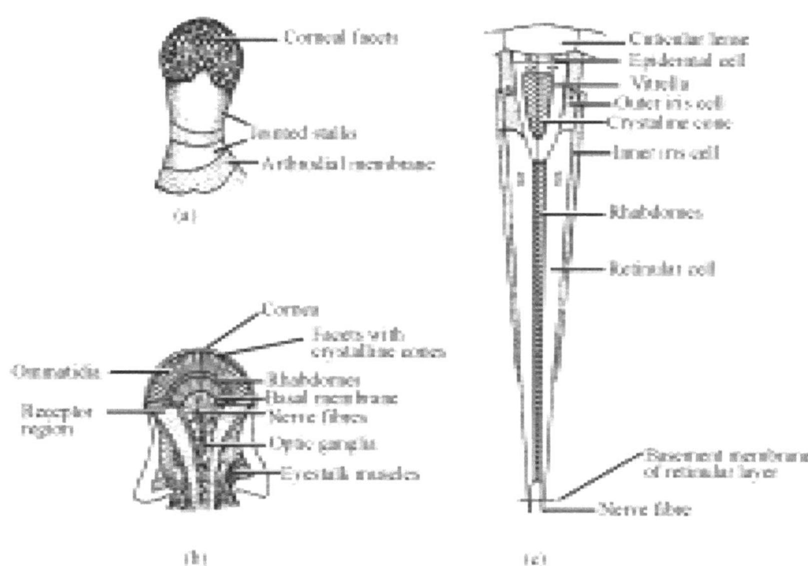

Fig. 8.7 : Structure of compound eye of prawn (a), longitudinal section of eye (b) and an ommatidium (c) of prawn.

The overall image is formed by combining the images from many individual ommatidium. This is known as mosaic vision. The images evoke nervous impulses, which are relayed to the cerebral ganglia. The eye is very sensitive to movement and has a wide field of vision often more than 180°. This is very advantageous to large predatory Malacostracans. Compound eyes can discriminate colour, initiating changes within skin cells to match the colour of the substratum. In crustaceans who are active nocturnally or in dark habitats, the ommatidia in the compound eye are typically organized to maximize light detection at the expense of visual acuity (superposition eyes), and multiple ommatidia combine to produce a single image. The eye in this condition of dark adaptation is said to be a superposition eye, while in bright light the eye with the curtain of pigments extending downwards is called an apposition eye.

Animals that are active in bright conditions generally have apposition eyes to maximize acuity with sensory input being received separately from each ommatidium. Apposition eyes are commonly found in crabs but shrimps and crayfish have superposition eyes. Some species can change from one eye system to another as light conditions appear by adjusting the position of screening pigment in the ommatidia. Many crabs have colour vision, and the overlapping visual fields of the two eyes indicate a capability of binocular vision. Crayfish have additional photoreceptors in their brain, which are presumably concerned with circadian rhythms.

The different refractive indices of water and air usually make vision difficult for focus to be achieved in both media. The flat lenses of the ommatidia of many crustaceans enable focus to be achieved in both the media, so that little change to the optics of the eye is necessary during initial emergence on to land. Normal light reflected from the near-by objects falls on the ommatidium and pass directly through the cuticular lens and crystalline cone to reach retinula. Due to the pigments in the iris cells, the light rays near to normal can reach it. Light rays other than normal are refracted towards retinula. So, each retinula receives rays from one small region of the object in view. In low light the iris cell pigments are withdrawn towards the surface so that rays received by one ommatidium can reach the retinulae of the nearby ones. A mosaic of overlapping images is produced with lack of clear vision. Objects may be better discerned in dim light than would be the case otherwise.

8.2.2. Statocytes

The informations on body movement, orientation and acceleration are obtained from statocytes. A pair of statocytes is present in most decapods. It is usually located inside the basal joint of the first antennae or at the base of the uropods or telson. Each statocyte is a sac, hair-lined pit, produced by an ingrowth of the ectoderm and has a crescentric opening fringed by large setae on the upper surface of the joint. The inside cuticle of the sac is thin and from it project several rows of feathery and sensory hairs upto the center of the cavity of the sac where the elongated nerve fiber is present. The hairs are in contact with the nerve fiber. The hairs arise from receptor cells in the sac wall. The receptor cells are supplied by a branch of the antennular nerve. Each statocyte contains a small statolith that mechanically stimulates the nerve fibers of the hairs as movements of the animal change its position. The statolith may be composed of fine sand particles bound together by the secretions from the statocyte wall. The nervous impulses produced pass to the cerebral ganglia and are relayed to produce the reflex actions needed for the decapod to maintain its normal equilibrium. The decapods also receive information regarding the position of their limbs and flexure of joints from suitably placed mechanoreceptors and internal proprioreceptors like stretch receptors. When ecdysis occurs, the cuticle and contents of the statocytes are shed. So, it is necessary for the decapods to replace the sand grains in the statocytes. The sand grain statoliths are replaced when the head is buried in sand or the animal inserts sand grains into the sac by means of chelae.

The sac is always open to the exterior, although in crabs and in some others the opening is reduced to a slit and is functionally closed. Complex statocytes are found in crabs with deep and irregular invaginations and may contain a labyrinth compartment and statocyte chamber. In such statocytes, these organs indicate gravitational position of the animal and also its position with regard to body movement. The rotation receptors develop special free sensory hairs that are not associated with the statolith and are stimulated by the motion of the fluid within statocyte chamber when animal moves. Its action mechanism is similar to semicircular canals of the vertebrates.

8.2.3. Chemoreceptors

Olfactory hairs or bristles or tactile setae, known as aesthetases are present generally over the external surfaces, antennae, food-gathering limbs, and mouthparts for chemoreception. They are also used to detect the predators. Chemoreceptors are also abundant on the tips of the walking legs where they are used to identify food in the substratum, while the mouthparts and chelae are very sensitive to dissolve chemicals and are helpful in sorting and selecting the food. Crabs can find food using chemical stimuli. The antennae have detectors that detect chemicals that stimulate the search for food. When similar detectors on the legs contact food, the cheliped quickly grasps the object and passes it into the mouth. Crab mouthparts have further receptors that are sensitive to particular chemicals. Crabs rely on a combination of these sense organs to find food, mates and predators.

8.2.4. Mechanoreceptors and other Minor receptors

There are other hair receptors, which are sensitive to touch, vibrations and sound, and air or water movements over the body. Many decapods produce sound by striking (percussion) or rasping (stridulating) or by internal mechanisms. Crabs have bristles and hairs, which act as touch receptors. These bristles occur all over the body but are most frequently found in bunches on the walking legs. These bristle's signal contact with a hard surface simply by bending, whilst other shorter hairs are sensitive to water current. Crabs can also hear and produce a variety of sounds. During courtship some species attract the female's attention by banging their cheliped on the ground or vibrating their walking legs. Each species has its own unique sound that can attract a female. Organs of sound reception in brachyuran crabs include the chordetonal organs on the hings of walking legs.

The specialized sound and vibration receptors include the antennal, calceoli of amphipods, the individual microstructures consisting of receiving elements arranged serially and attached to the antennal segment by a slender stalk. In more advanced decapods, the basal elements are expanded into a cup-like receptacle and the stalk is distally expanded into a bulla or resonator.

In decapods, the tactile organs are in the form of some plumose setae, fringing the flattened portions of the appendages like the rami of the pleopods. The setae are hollow cuticular outgrowths supplied with nerve fibers, and consist of two segments. The basal segment or the shaft is slightly swollen and attached to the integument by a membrane. The blade or distal segment bears two rows of small barbs.

The inner smaller branch of the outer feeler of each antennule of shrimps bears a longitudinal groove containing numerous olfactory setae. These setae consist of a basal shaft that is attached to the integument by a flexible membrane, and a distal blade segment is bluntly rounded. A small nerve fiber from the olfactory branch of the antennulary nerve innervates each seta in all decapods.

CHAPTER - 9

Glossary

Antenna : The long segmented appendages located behind the eyestalks. These allow the shellfish to interact with its environment by touch and chemoreception.

Acron : At the front (or anterior end) of the shellfish body there is an unsegmented or presegmented region called acron.

Antennule : Shorter segmented appendages located between and below the eyestalks, sensory organs; these also use chemoreception to "smell" and "taste".

Appendages : Ten legs (five pairs) including a claw-bearing pair with spines used for feeding and defense, followed by three pairs of sharply pointed walking legs, and a pair modified as flat swimming paddles at the rear, swimming legs.

Appetite : It is the amount of food eaten voluntarily at one time, appears to be the increase of stomach fullness, although this does not explain the entire appetite phenomenon.

Apron : Abdomen of a crab, which is folded under the body; male's is shaped like the Washington Monument or an inverted Y. An immature female's is triangular (pyramid shaped) and mature female's is semicircular, like the dome of the Capitol building.

Aquaporin : This is a protein present in the membranes of the renal collecting duct cells.

Autotomy : To break the leg spontaneously to escape.

Bilirubin and biliverdin : Bile contains the fat emulsifying bile salts along with bile pigments, biliverdin and bilirubin, which arises from the breakdown of red blood cells and hemoglobin.

Carapace : The shell covering the body provides rigidity and protective covering. It is made of chitin and is the part of exoskeleton (hard outer covering) that covers the head and thorax (center) of the shellfish.

Cheliped : The first pair of legs, carries the large claw which is used for defense and obtaining food. In crabs male's claws are blue tipped with red; females are red.

Chromatolysis : During the degeneration of the cell body of the neuron, disorganization of the Nissl's bodies occur by a process called chromatolysis.

Chyle : Fat absorbed from intestinal mucosa.

Coagulogen : It is the clotting factor in granular cells of decapods.

Cor frontale **:** The *cor frontale* or the accessory heart of shellfish.

Diastole : The expansion of heart.

Endoneurium : It consists of a thin layer of loose connective tissue surrounding individual nerve fiber.

Epineurium : It is the dense connective tissue layer with many blood vessels that surrounds the nerve.

Forage ratio : The selectivity of food can be studies by calculating an index called "forage ratio" or "available factor".

Furca : The telson may bear two processes, or rami, which together form the furca.

Gill rakers : The inner border of the fish gill arch extends to form gill rakers.

Glycocalyx : Negatively - charged coat of foot processes of podicytes of Bowman's capsule.

Hastate ossicle : A large triangular ossicle is embedded in the middle of the floor of the cardiac stomach in shellfish. It is called the hastate ossicle.

Hematopoisis : Formation of blood cells (hematopoisis) occurs in many organs of fish.

Hepatopancreas (midgut gland) : Extremely large organ with several functions, including the secretion of digestive enzymes and absorption and storage of digested food. It fills most of the area around the stomach, depending on its contents of food reserves and water.

Heterocercal : The heterocercal (unequal) tail is characteristic of chondrichthyes and some primitive bony fishes.

Holonephros : The kidney is holonephros when it extends to the entire length of the body. Such kidney is found in larvae of some cyclostomes.

Homocercal : The homocercal (equal tail) is characteristics of higher bony fishes.

Index of selectivity : It is a measure to compare a particular item of food found in the gut and that of in the environment.

Kopfdarm : The head gut or Kopfdarm is often considered in terms of its two components, the oral (buccal) and gill (branchial or pharyngeal) cavities.

Lateral spines : Paired points on the widest outside edges of the carapace in shellfish.

Leptomeninx : The choroid plexus, consisting of layers of mesodermal tissues, is a highly vascularized area in holosteans and elasmobranchs. This is also known as leptomeninx.

Luciferase : This enzyme oxidizes luciferin to oxyluciferin that emits light for bioluminescence.

Luciform : It is an indole derivative containing tryptamine, arginine and isoleucine.

Mauthner chiasma : The Mauthner axon forms Mauthner chiasma crossing each other.

Mauthner neuron : Actinopterygii (bony fishes) contain a pair of very large multipolar neuron, known as Mauthner neuron in the medulla at the level of cranial nerve VIII, which is absent in elasmobranchs (except at the young stage).

Mesonephros : The kidney is mesonephric when the middle region functional. The mesonephros serves as the main excretory organ for aquatic vertebrates like fish.

Methyl farnesoate (MF) : It is a crustacean hormone.

Myogenic : Most fish hearts are myogenic i.e no neural input from the brain is necessary for each heart beat and show a complex electromyogenic wave-form.

Neuromast : The sensory cells gather at different points to form a sense hillock or neuromast.

Ophisthonephros : In fish the most anterior tubules have been lost, some middle tubules are associated with testes and there is a multiplication of tubules posteriorly. Such kidney is known as ophisthonephros.

Otolith : The sand particles of gelatinous cupula of the hair cell act as otolith.

Perineurium : Epineurium penetrates the nerve to form the perineurium that surrounds bundles of nerve fibers.

Pronephros : Generally in fishes the tubules of the anterior region become functional in early life and such type of kidney is designated as pronephros.

Protocercal : The protocercal (first tail) is the most primitive one. The hind end of the notochord or the vertebral column is straight or divides the caudal fin into two equal lobes, epichordal (dorsal) and hypochordal (ventral) lobes.

Proventriculus : The proventriculus (foregut) contains a triturating gastric mill in all decapods.

Pylorus : It is a muscular sphincter which controls the transport of food from the stomach into midgut.

Rugae : Intestinal transverse folds (rugae).

Rumpfdarm : The foregut and hindgut together are known as Rumpfdarm.

Spawning fast : Due to gonadal maturation, feeding decreases in mature fish, which is termed as "spawning fast".

Sponge : Egg masses. Numbers of eggs vary. Some decapods (crabs) may contain as many as 8,000,000. They are attached to swimmerets.

Swimmerets (pleopods) : Paired abdominal appendages under the apron of the female crab on which the eggs are carried until they hatch.

Systole : The contraction heart.

Typhosole : The absorption capacity of the intestinal area is increased by making its walls into lengthwise folds (typhosole).

Villi : Intestinal finger-like projections (villi).

Vitellogenesis : Accumulation of vitellin or yolk protein.

Walking legs : Used for movement; crabs are capable of walking forward or diagonally, but usually they walk sideways.

CHAPTER - 10

Bibliography

Ahyong, S.T. & O'Meally, D.2004. Phylogeny of the decapoda Reptantia: resolution using three molecular loci and morphology-Raffles Bull.zool, 52:673-693.

Alexander, R.M.1967. Functional design in fishes. Hutchinson Lib, London, 160pp.

Al-Mohann, S.Y. and Nott, J.A. 1987.R-cell and digestive cycle in *penaeus semisulcatus* (crustaceans: Decapoda). Marine Biology, 95: 129-137.

Armstrong, D.A.;Strange,K.;Crowe,J.;Knight, A. and Simmons,M.1981. High salinity acclimation by the prawn, *macrobrachium rosenbergii*, uptake of exogenous ammonia and changes in endogenous nitrogen compounds. Biol. Bull. Mar. Biol. Lab. Woods Hole. 160:349-365.

Aronson, L.R. 1970. Functional evolution of the forebrain in lower vertebrates. In: Development and evolution of behaviour. Edited by L.R.Aronson *et al.* Freeman Press. San Francisco: 75-107.

Augusto, A; Green, L.J.; Laure, H.J. and McNamara, J.C.2007. Adaptive shifts in osmoregulatory strategy and the invasion of freshwater by branchyuran crabs: Evidence from *Dilocarcinus pagei*. J.Exp.Zool.307A:688-698.

Bagarinao, T. 1994. The natural life history of milk fish. SEAFDEC Asian Aquaculture.XVI (3).Sep,94:3-4.

Bagenal,T.B. 1978. Aspects of fish fecundity. *In*: Ecology of Freshwater Fish Production. Edited by Gerking, S.D.Wiley, New York: 75-101.

Barlow, J. and Ridgway, G.J.1969. Changes in serum protein during the molt and reproductive cycles of the American lobster, *Homarus americanus*. J.Fish.Res.Board Can.26:2101-2109.

Barnes, R.D. 1982. Invertebrate Zoology. 4th Edition. Holt-Saunders Int. Edition. 1089pp.

Bauer, R.T. 1977. Antifouling adaptations of marine shrimps. Functional morphology and adaptive significance of a antennular preening by the third maxillipeds. Mar.Biol.40:261-276.

Berg,T. and Steen, J.B. 1965. Physiological mechanisms for aerial respiration in the eel. Comp. Biochem. Physiol. 15:469-484.

Biswas, S.P. 2002. Fundamentals of Ichthyology. Narendra Pub.House, New Delhi. 392pp.

Bohle, A. and Walvig, F. 1964. Klin. Wochschr.42:415-421.

Bone,Q. 1963. The Central Nervous System. *In*: Biology of Myxine.Edited by Brodal,A. and Fange,R. Oslo Univ. Press, Oslo:50-91.

Bone,Q.;Marshall, N.B. and Blaxter, J.H.S. 1995. Biology of Fishes. 2nd Edition. Chapman and Hall, Blackie Academic and Professionals. U.K. .

Brethes, J.C., Parent, B.and Pellerin,J. 1994. Enzymatic activity as an index of trophic resource utilization by snow crab *chinoecetes opilio* (*O.Fabricius*). Journal of crustacean Biology, 14: 220-225.

Brown, F.A.2002. Invertebrates. 1st Edition Biotech Books, Delhi. 507pp.

Buddington, R.K. and Doroshov, S.I. 1986. Development of digestive secretions in white sturgeon juveniles (*Acipenser transmontanus*). Comp. Biochem. Physiol. 83A:233-238.

Bulger,R.E. and Trump, B.F. 1968. Renal morphology of the English sole, *Parophrys vetulus*. Am.J.Anat. 123:195-225.

Bullock, T.H. and Horrldge, G.A. 1965. Structure and function in the nervous systems of invertebrates. Freeman, San Francisco.

Bullock, T.H.;Bodznick, D.A. and North cutt, R.G. 1983. The phylogenetic distribution of electroreception: Evidence for convergent evolution of a primitive vertebrate sense modality. Brain Research Reviews. 6:25-46.

Busselen, P. 1970. Effect of molting cycle and nutritional conditions on haemolymph proteins in *Carcinus maemas*. Comp.Biochem. Physiol. 37:73-83.

Calman,W.T. 1909. Crustacea *In*:A Treatise on Zoology Lankester, E.R. (ed.). London:Adam and Black:346.

Cameron,J.N.1975. Morphometric and flow indicator studies of the teleost heart.Can.J.Zool.53:691-698.

Catlett,R.H. and Millich,D.R. 1976. Intracellular and extracellular osmoregulation of temperature - acclimated goldfish, *Carassius auratus* L. Comp. Biochem. Physiol.55A:261-269.

Ceccaldi, H.J.1982.Contribution of physiology and biochemistry to progress in aqua culture. Bull.Japan.Soc.Sci.Fish.48:1011-1028.

Ceccaldi, H.J. 1989. Anatomy and Physiology of digestive tract of crustaceans decapods reared in aquaculture. Adv. Trop.Aquacult.Tahti, Aquacop.IFREMER.Acies de colloque 9:243-259.

Ceccaldi, H.J. 1997. Anatomy and physiology of the digestive system in Crustacean nutrition. The Aquaculture world society. Baton Rouge D'Abramo, L.R., Conklin D.E. and Akiyama, D.M. (eds.):261-291.

Chang, E.S. 1995. Physiological and biochemical changes during the molt cycle in decapod crustaceans: An overview. J.Exp.Mar.Biol.193:1-14.

Chapmon,G. and Barker, W.B. 1964. Zoology. 1st Edition. Longman, London:144-168.

Charmantier, G.;Soyez,C. and Aquacop. 1994. Effect of molt stage and hypoxia on osmoregulatory capacity in the peneid shrimp, *Penaeus Vannamei*, J.Exp.Mar.Biol.Eco.178:233-246.

Chatwal, M.S. 2000. A Textbook of Invertebrates. 1st Edition. Campus Books Int., New Delhi: 440pp.

Chieffi, G.1962.Endocrine aspects of reproduction in elasmobranch fishes. Gen.Comp.Endocrinol.Suppl.1:275-285.

Ciofi, M. 1984. Comparative ultrastructure of arthropod transporting epithelia. Am.Zool.24:139-156.

Cobb III, B.F.1978. Biochemistry and Physiology of shrimp - Effect on use as Food.Journal Article No.-12733 of the Texas Agricultural Experiment Station:141-164.

Conte, F.P. 1969. Salt Secretion.*In*:Fish Physiology.Edited by Hoar,W.S. and Rondall, D.J.Academic Press,New York Vol.I:241-292.

Craik, J.C.A. 1978.Kinetic studies of vitellogenin metabolism in the elasmobranch, *Scyliorhinus canicula*, L.Comp.Biochem.Physiol.A.61A:355-361.

Crompton, V.;Bergheim,A.;Gausen,M.;Naess,A.and Holland,P.M. 2003.Effect of low oxygen on fish performance. www.ewos.com .

Dall, W., and Moriarty,D.M. 1983. Functional aspects of nutrition and digestion in internal anatomy and physiology regulation *In*:The Biology of Crustacea.Mantel,L.H.(ed.), Vol-5, Academic press, New York.

Dall, W.1992 Feeding, Digestion and assimilation in penaeidae in proceedings of aquaculture nutrition workshop (G.L.Allan and W.Dall,eds.), Salamander Bay,15-17 April 1991 NSW Fisheries, Salamander Bay:57-63.

De Kleijn, D.P.V., and Van Herp, F.1998. Involvement of the Hyperglycemic neurohormone family in the control of reproduction in decapod, crusteceans. Invert.Reprod.Dev.33:263-272.

Demaski,L.S.,Bauer,D.H. and Gerald,J.W.1975.Sperm release evoked by electrical stimulation of the fish brain, a functional anatomical study.J.Exp.Zool.191:215-221.

Digkgraaf, S.1962.Zeitschr.Vergl.Physiol.27:587-605.

Duttamunshi, J.S. 1960. Ind. J. Zootomy.1(3):135-174.

Dutta,H.M. and Hossain,A.M. 1993. Pattern of caecal distribution in teleost fish.*In*:Advances in Fish Research.Vol.I.Edited by B.R.Singh. Narendra Publishing House. New Delhi:267-280.

Eisen, A.Z.; Henderson, K.O.;Jeffrey, J.J. and Bradshaw, R.A. 1973. Acdlagenolytic protease from the hepatopancreas of the fiddler crab *Uca pugilator*. Purification and properties. Biochemistry. 12:1874-1922.

Farrell, A.P.;Hammons,A.M.;Graham, M.S. and Tibbits, G.F. 1998. Cardiac growth in rainbow trout, *Oncorhynchus mykiss*.Can.J.Zool.66:2368-2373.

Felder, D.L. & Felgenhauer, B.E. 1993. Morphology of the mid gut - Hindgut junature in the ghost shrimp *Lepidophthalmus lousiunensis* (Schmitl). (Crustacea : Decapoda : Thalassinidae)-Acta zool, 74:263-276.

Felgenhauer, B.E. and Abele, L.G. 1983. Phylogenetic relationships among shrimp like decapods. *In*: Crustacean Phylogeny. Crustacean Issues Schram, F.R.(ed). Vol.1 Rotterdam, Netherlands:Baikema: 291-311.

Felgenhauer, B.E. 1992. Internal Anatomy of the Decapoda:An Overview. In:Microscopic anatomy of Invertebrates, : Decapoda Crustacea, Wiley - Liss, Inc.Vol - 10 : 45 - 75.

Figueiredo, M.S.R.B.;Kricker, J.A. and Anderson, A.J. 2001. Digestive enzyme activities in the alimentary tract of redclaw crayfish, *Cherax quadricarinatus* (Decapoda : parastacidae), Journal of Crustacean Biology , 21:334-344.

Florkin, M. 1962. La re'gulation isosmotique intracellular chez les inverte'bre's marins euryhalins. Bull. Acad.Roy.Belg.Clin.Sci.48:687-694.

Fortaine, C.T., and Lightner, D.V. 1973. Observations on the process of wound repair in penaeid shrimp. J.Invert.Pathol. 22:22-33.

Foster, C.A.; and House, H.D. 1978. A morphological study on gills of the brown shrimp *Penaeus aztecus*. Tissue cell 10:77-92.

Foster, C.;Amado,E.M.;Souza,M.M and Freire, C.A.2010. Do osmoregulators have lower capacity of muscle water regulation than osmoconformers. A study on decapods crustaceans. J. Exp. Zool. 313A:80-94.

Freire, C.A.; Onken, H. and Mc Namara, J.C. 2008. A structure – function analysis of ion transport in crustacean gills and excretory organs. Comp. Biochem. Physiol. 151A:272-304.

Garnaud,T. 1950. As cited by Khanna, S.S. 1992.

Gibson, R., and Barker, P.L. 1979. The decapod hepatopancreas oceanogr. Mar. Biol. 17:285-346.

Gonzalez-Pena, M;Anderson, A.J.;Smith, D.A. and Moreira, G.S.2002. Effect of dietary cellulose on digestion in the prawn *Macrobrachium rosenbergii.* Aquaculture, 211:291-303.

Goodman, S.H., and Cavey, M.J., 1990. Organisation of a phyllobranchiate gill from the green shore crab *Carcinus maenas* (crustacean, Decapoda), Cell Tissue Res. 260:495-505.

Graham, M.S. and Farrell, A.P. 1989. The effect of temperature and adrenaline in the performance of a perfused trout heart Physiol. Zool. 62:38-61.

Graham, M.S. and Farrell, A.P. 1992. Environmental influences on cardiovascular variations in rainbow trout. *Oncorhynchus mykiss.* J.Fish Biol. 41:851-858.

Groat, C, Margolis, L. and Clarke, 1995. Physiology, Ecology of Pacific Salmon. Department of Fisheries and oceans, Biological sciences Branch Pacific biological station Nanaime, British Columbia, Canada.

Gruber, S.H.; Gulley, R.L. and Brandon, J. 1975. Duplex retina in seven elasmobranch species. Bull. Mar. Sci.25:353-358.

Guillaume, J.1997. Protein and Aminoacids in crustacean nutrition. In: The Aquaculture world society Baton Rouge, (D'abramo, L.R; Conklin, D.E. and Akiyama, D.M.(eds):26-50.

Guillame, J. and Ceccaldi, H.J. 2001. Digestive physiology of shrimp *In*: Nutrition and feeding of fish and crustaceans. Guillaume,J.; Kaushik, S.;Bergot, P.and Metailler, R.(eds.). Praxis Publishing Ltd., Chichester;239-252.

Guraya, S.S. 1993. Follicular (or oocyte) atresia and its causes and functional significance in fish ovary. *In*: Advances in Fish Research. Edited by B.R. Singh. Vol.1. Narendra Publishing House. New Delhi, pp. 313-332.

Hangerman, L.1983. Haemocyanin concentration of juvenile lobsters, *Homarus gammarus* in relation to moulting cycle and feeding conditions. Mar. Biol.77:11-17.

Hanumantha Rao, 1974. As cited by Khanna, S.S., 1992.

Hasler, A.D. and Scholz, A.T. 1983. Olfactory imprinting and homing in salmon:Investigations into the mechanism of the imprinting process. Zoophysiol. 14, Springer-Verlag, Berlin. 134pp.

Hertzler, P.L.2005. Cleavage and gastrulation in the shrimp *Penacus(Litopenaeus) Vannamei* (Malacostraca, Decapoda, Dendrobranchiata) Arthropod Structure & Development. 34:455-469.

Hibiya, T.1982. An atlas of fish histology, normal and pathological features. Lodansha Ltd. Gustav . Fisher Verlag. New York.

Hickman, C.P.Jr. and Trump, B.F.1969. *In*:Fish Physiology. Edited by Hoar, W.S. and Randall, D.J.Academic Press, New York. Vol-I:91 -239.

Hill, 1976. As cited by Srivastava, C.B.L. 1999. Fish Biology. Narendra Pub.House.

Hoar, W.S. and Randall, D.J.(Eds).1970. Fish Physiology Vol-IV. Academic Press, New York. 494pp.

Hoar, W.S. 1975. General and Comparative Physiology. 2nd. ed. Prentice-Hall, Englewood Cliffs, N.J.848pp.

Holwerda, D.A. and Vonk, H.J.1973. Emulsifiers in the intestinal juice of crustacea - Isolation and nature of surface - active substances from *Astacus leptodactylus,* Esch. And *Homarus Vulgaris.* Comp.Biochem. Physical. 458:51-58.

Hollman, J.D. & Hand, S.C. 2009 Metabolic depression is delayed and mitochondria impairment averted during prolonged anoxia in the ghost shrimp, *Lepidophthalmus louisianensis* (Schmitt, 1935), J. exp. Mar Biol Ecol.376:85-93.

Hoppkin, S.P. and Nott, J.A. 1980.Studies on the digestive cycle of the shore crab *Carcinus maenas*(L.) with special reference to the B cells in the hepatopancrease. J.Mar.Biol. Assoc, U.K. 60:891-907.

Hornung, D. and Stevenson, J.1971. Charges in the rate of chitin synthesis during the crayfish molting cycle. Comp.Biochem. Physiol.40B:341-346.

Hughes, G.M. and Grimstone, A.V.1965. The fine structure of the secondary lamella of the gills of *Gadus pollachius.* Quart. J. Micro. Sci. 106:343-353.

Jaffe, E.R.1964. Metabolic processes involved in the formation and reduction of methemoglobin in human erythrocytes. *In*: The Red Blood Cell. Edited by Biship, C. and Surgemor, D.Academic Press, New York. 397-422.

Johnson, P.T. 1980 . Histology of the Blue Crab, *Callinectus Sapidus*: A model for the Decapoda, New York : Praeger.

Jordan, J. 1976. The influence of body weight on gas exchange in the air breathing fish *Clarius batrachus.* Comp. Biochem, Physiol. 534:304-310

Joshi and Khanna, 1981. As cited by Khanna, S.S.

Kalmijn, A.J.1971. The electric sense of sharks and rays. J.Ex.Biol.55:371-383.

Kalmijn, A.J. 1978. Experimental evidence of geomagnetic orientation in elasmobranch fishes. *In*:Animal migration, Navigation and homing. Edited by Schmidt-Koenig, K. and Keeton, W.T. Springer - Verlag, Berlin:347-353.

Kapoor, B.G.; Smit, H. and Verighina, I.A.1975. The alimentary canal and digestion in teleosts. Adv. Mar. Biol.13:109-239.

Karnaky, K.J. Jr. 1986. Structure and function of the chloride cell of *Fundulus heteroclitus* and other teleosts. Amer. Zool. 26:209-224.

Katayama, T.; Katama, T. and Chichester, C.O.1972. The biosynthesis of astaxanthin. VI. The carotenoids in the prawn *Peneaus japonicas* Bate (Part II). Int. J. Biochem. 3:363.

Kavaliers, M.1979. Pineal involvement in the control of circadian rhythmicity in the lake chub, *Couesius plumbens.* J.Exp.Zool.209:33-40.

Kempton, R.T.1956. The problem of the special segment of the elasmobranch kidney tubule. Year Book Am. Phil.Soc . 210-212.

Kerr, J.G.1919. Textbook of Embryology Vol-II.Macmillan, New York.

Khanna, S.S. and Singh, H.R. 1965. Proc. Zool. Seminar, Res Trends in Animal Morphology, 187-194.

Khanna, S.S. and Singh, H.R. 1967. On the morphological peculiarities of the valvula cerebella in some teleosts. Zool. Anz-178:413-419.

Khanna, SS.1992. An Introduction to Fishes. Central Book Depot, Allahabad, India. pp-530.

Kleve, MG, yudin, A.I. and clark, W.H.1980. Fine structure of the unistellate sperm of the shrimp *Sicyonia ingentis*(Natantia). Tissue cell 12:29-45.

Kotpal, R.L.2001 Arthropada. Rastogi Publications . New Delhi p – 17-71.

Kramer, D.L. and McClure, M.1982. Aquatic surface respiration, a widespread adaptation to hypoxia in tropical freshwater fishes. Env.Biol. Fish . 7:47-55.

Kuchnow, K.P.1971. The elasmobranch papillary response. Vision Res.11:1395-1406.

Kumar, S. and Singh, M.P.1990 . Histological and histochemical studies of the *Auerbach plexus* of colon of rabbit, *Oryctolagus cunniculus*. Ind. J.Z. Spect.1:31-35.

Kumar, S. and Tembhre, M.1998. Anatomy and Physiology of Fishes. Vikas Publishing House Pvt. Ltd., pp . 274.

Kuramoto,T. and Ebara, A.1984. Neurohormonal modulation of the cardiac out flow through the cardioarterial valve in the lobster. J.Exp.Biol. 111:123-130.

Kurian, A.1977.Index of relative importance – a new method for assessing the food habits of fishes. Indian J.Fish.24(1 & 2):217-219.

Lagler, K.F.; Bardach, J.E.; Miller, R.R. and Passino, D.R.M. 2003. Ichthyology. 2nd Edition. John Wiley and Sons.Inc, Canada.

Leblanc, L.A.2002. Observations on reproductive behaviour and morphology in the ghost shrimp *Lepidophthalmus louisianensis* (Schmitt, 1935) Decapoda: Thalassinidae: Callianassidae: 1-47.

Lenfant, C. and Johansen, K.1996. Respiratory function in the elasmobranch *Squalus suckleyi* G.Respir. Physiol. 1:13-29

Mangum, C.P.;McMahon, B.R.;deFur, P.L. and Wheatly, M.G.1985. Gas exchange, acidbase balance and the oxygen supply to tissue during a molt of the blue crab, *Callinectes sapidus*. J.Crustac. Biol . 5:188-206.

Mantel, L.; Bliss, D.;Sheehan, S. and Martinez, E.1975. Physiology of hemolymph, gut fluid and hepatopancreas of the land crab, *Gecarcinus lateralis* in various neuroendocrine states. Comp.Biochem.Physiol.51A:663-671.

Mantel, L.H., and Farmer, L.L.1983. Osmotic and ionic regulation. *In* – The Biology of crustacean, Bliss, D.E. and Mantel, L.H.(eds), 5:53-161.

Martin, J.W. and Davis, G.E.2001. An updated classification of the recent crustacea- Natural Science Series:History Musium of Los Angeles Country. 39:1-132.

McNamara, J.C.;Rosa, J.C.;Greene, L.J. and Augusto, A.2004.Free amino acid pools as effectors of osmotic adjustment in different tissues of the fresh water shrimp, *Macrobrachium olfersii* (Crustacea, Decapoda) during long term salinity acclimation. Mar.Freshwat.Behav.Physiol.37:193-208.

Mercaldo-Allen, R.1991. Changes in the blood chemistry of the American lobster, *Homarus americanus* over the molt cycle. J.Shellfish Res.10:147-156.

Morris, S.2001. Neuroendocrine regulation of osmoregulation and the evolution of air-breathing in decapods crustaceans. J.Exp.Biol.204(Pt-5):979-989.

Motah, H.1981. Studies on the biology of the tiger prawn, *Penaeus monodon* in the Philippines. Technical Report. Tigbauan, SEAFDEC Aquaculture Department. 128pp.

Moyle, P.B. and Cech, J.J.Jr. 1988. Fishes: An Introduction to Ichthyology. 2nd ed. Prentice – Hall, Inc, N.J. 559pp.

Murphy, G.I.1968.Pattern in life history and the environment. Amer. Nat.102:390-404.

Mussarelli, R.A.A.1977. Chitin. Pergamon Press Ed. Oxford.

Nieuwenhuys, R. 1964. Comparative anatomy of the actinopterygian forebrain. J. Hirnforsch.6:171-192.

Nieuwenhuys, R. 1966. The interpretation of the cell masses in the teleostean forebrain. *In* : Evolution of forebrain, phylogenesis and ontogenesis of the forebrain. Edited by R. Hasler and G. Stephan. Gerg Thieme Verlag, Stuttgart: 32-39.

Ogawa, 1961. As cited by Srivastava, C.B.L.1999. Fish Biology.Narendra Publishing House, New Delhi.327pp.

Omondi, J.G. and Stark, J.R. 1995. Some digestive carbohydrases from the midgut gland of *Penaeus indicus* and *Penaeus vennamei* (Decapoda:penaeidae). Aquaculture, 134:121-135.

O'Halloran, M.J. and Idler, D.R.1970. Identification and distribution of the Leydig cell homology in the testis of sexually mature Atlantic salmon, *Salmo salar*.Gen.comp. Endocrinol.15:361-364.

Page, J.W. 1976. *et al.* Hydrogen ion concentration in the gastrointestinal tract of channel catfish. J.Fish Biol. 8:225-228.

Pal, A.K.; Dasgupta, S. and Sahu, N.P. 2011. Fish Physiology. *In*:Handbook of Fisheries and Aquaculture. Edited by Ayyappan, A.; Moza, U.;Gopalkrishna, A.;Meena kumari, B.;Jena, J.K. and Pandey, A.K. Indian Council for Agric.Res., New Delhi:658-704.

Parker, T.J. and Haswell, W.A.1967. A Text Book of Zoology, MacMillan & Co, London (Reprint).952pp.

Pawley, A. 2004. Are crustaceans shell fish? A whiff of scandal in English lexicography. June 2004. [accessed 2007 April 30].

Peterson, D.R. and Loizzi, R.F. 1974. Ultrastructure of the crayfish Kidney –coelomosac, labyrinth, nephridial canal. J.Morph.142(3):241-263.

Pe'queux, A. and Gills, R. 1981. Na^+ fluxes across isolated perfused gills of the Chinese crab, *Eriocheir sinensis*. J.Exp.Biol. 92:173-186.

Pe'queux, A. 1995. Osmotic regulation in crustaceans. J.Crust. Biol. 15:1-60.

Prasser, C.L. 1973. Comparative Animal Physiology. Philadelphia, WB.Saunders Co.

Pressley, T.A.; Graves, J.S. and Krall, A.R. 1981. Arniloride sensitive ammonium and sodium transport in the blue crab. Am. J.Physiol 241(Regulatory Integrative Comp.Physiol.) 10:R370 - R378.

Regaud, C. and Policard, A.1902. Etude sur le tube urinifere de la lamproie. Compt. Rend. Assoc. Anat.4:245-261.

Regnault, M. 1984. Salinity-induced changes in ammonia excretion rate of the shrimp, *Crangon crangon* over a winter tidal cycle. Mar. Eco. Pro. Series. 20:119-125.

Riegel, J.A. 1972. Comparative physiology or Renal Excretion. Edinburgh:Qliver and Boyd.

Romer and Parson, 1977. As cited by Srivastava, C.B.L.1999. Fish Biology. Narendra Publishing House, New Delhi. 327pp.

Root, R.1931. The respiratory fuction of the blood of marine fishes. Biol. Bull. 61:427-456.

Royce, W.F. 1972. Introduction to the Fishery Sciences. Academic Press, New York. 351pp.

Saffran, W.A. and Gibson, Q.H. 1976. Kinetics of the Bohr effect of menhaden haemoglobin, *Brevoortia tyrannus*. Biochem. Biophys. Res. Commun. 69:383 - 388.

Sakai, K.2005. Callianassoidae of the world (Decapoda, Thalassinidae)- crustacean Monogr, 4:1- 285.

Salmon, M. 1983. Courtship, mating systems and sexual selection in decapods. *In*:Studies in Adaptation:The Behaviour of Higher Crustacea Edited by Rebach, S. and Dunham, D. John Wiley & Sons, Inc, Canada:143 - 169.

Salmon, M. and Hyatt, G.W. 1983. Communication. *In*:The Biology of Crustacea:Behaviour and Ecology. Edited by Vernberg, F.I. and Vernberg, W.B.Academic Press, New York:1-40.

Samantaray, K. 2012. Principles of Biochemistry with Special Reference to Fishes. Narendra Publishing House, New Delhi. pp.360.

Schneider, S. 1903 Ein Beitrag zur Kenntre's der Physiologie der Niereniederer Wirbeltiere. Skand. Arch. Physiol.14:383-389.

Schoffeniel, E. and Gills, R. 1970. Osmoregulation in Aquatic Arthropods. *In*:Chemical Zoology. Edited by Florkin, M. and Scheer, B.T. Vol. 5 (Pt-A). Academic Press, New York.pp255 - 286.

Scholtz, G. 2004. Evolutionary Developmental Biology of Crustacea. Crustacean Issues 15.A.A. Balkema Publishers:43 - 91.

Schram, F.R. and von Vaupelklein, J.C.(Eds)-2012. Treatise on zoology - Anatomy, Taxonomy, Biology:The Crustacea Vol-9(B) Koninklijke Brill N.V., Boston.

Shrivastava, V.M.S., Shrivastava, D.K. and Tripathi, R.S. 1980. Morphological and histochemical studies of the pineal organ in *Wallago attu*. Int. J. Acad. Ichthyol. 1:31-34.

Singh, H.R. 1971. On the thalamic and hypothalamic nuclear areas of some teleosts. Acta. Anat. 78:574-590.

Singh, W 1981. Proc.Nat.Acad.Sci.51B:197-200.

Singh, H.R. 1993. Fish brain, a fascinating structure. *In*: Advances in fish Research. Edited by B.R. Singh. Vol 1. Narendra Publishing House, New Delhi:281-294.

Smirnova, L.I. 1966. Digestive leucocytosis of bream (*Abramis brama*). *In*: Biology of Fishes of the Volga reservoirs. Tr.Inst.Biol.Vnutr.Vod./ Trans. Inst. Biol. Inland waters. 10(3):143-147.

Smith, 1960. As cited by Srivastava, C.B.L. 1999. Fish Biology, Narendra Publishing House, New Delhi. 327 pp.

Smith, B.W. and Lovell, R.T. 1973. Determination of apparent protein digestibility in feeds for channel catfish. Trans. Am. Fish. Soc.102(4):831 - 835.

Spaargaren, D.H.; Richard, P. and Ceccaldi, H.I. 1982. Excretion of nitrogenous products by *Penaeus japonicus*, Bate, in relation to environmental osmotic conditions. Comp. Biochem. Physiol. 72A: 673 - 678.

Spindler - Barth, M. 1976. Changes in the chemical composition of the common shore crab, *Carcinus maenas* during the molting cycle. J. Comp. Physiol. 105:197 -205.

Srivastava, C.B.L. 1999. Fish Biology. Narendra Publishing House, New Delhi 327pp.

Stanzel, C. & Finelli, C.2004. The effects of temperature and salinity on ventilation behaviour of two species of ghost shrimp (Thalassinidea) from the northern Gulf of mexico : a laboratory study; J.exp. Mar Biol. Ecal., 228:197 - 308.

Steinacker, A.1978. The anatomy of the decapods crustacean auxiliary heart. Biol. Bull. 154:497 - 507.

Strange, K. 20004. Cellular volume homeostasis. Rev.Adv.Physiol.Edu . 28:155 - 159.

Stutman, L.J. and Dolliver, M. 1968. Mechanism of coagulation in *Gecarcinus lateralis*.Am.Zool.8:481 - 489.

Tazaki, K. and Cooke, I.M. 1979. Spontaneous electrical activity and interaction of large and small cells in the cardiac ganglion of the crab, *Portunus sanguinolentus*. J. Neurophysiol. 42:975-999.

Telford, M. 1968. Changes in the blood sugar compostion during the molt cycle of the lobster, *Homarus americanus*. Comp. Biochem. Physiol. 26:917-926.

Tembhre, M. and Kumar, S. 1995. Acetylcholinesterase activity and enzyme kinetics in the gut of *Cyprinus carpio* subjected to acute and chronic exposures to copper. J. of Ecobial. 7:121 - 124.

Thomas, P.C.; Rath, S.C. and Das Mohapatra, K.2003. Breeding and Seed Production of Finfish and Shellfish. Daya Publishing House, New Delhi 402pp.

Towle, D.W. and Kays, W.T. 1986. Basolateral localization of $Na^{+}K$-ATPase in gill epithelium of two osmoregulating crabs, *callinectes sapidus* and *carcinus maenas*. J.Exp. Zool. 239:311-318.

Utida, S. and Hirano, T.1973. Effects of changes in environmental salinity on salt and water movement in the intestine and gills of the eel *Anguilla japonica*. *In*:Responses of Fish to Environmental changes. Edited by Chavin, C.W.Springfield, III, Chas. C.Thomas:240-278.

Vacca, L.L. and Fingerman, M. 1983. The roles of haemocytes in tanning during the molt cycle:A histochemical study of the fiddler crab, *Uca pugilator*. Biol. Bull, 165:759-777.

Volya, G. 1996. Some data on digestive enzymes in some Black Sea fishes and a micromodification of a method for the identification of a trypsin, amylase and lipase. *In*:Fisiologia morskikhyzhitvotnykh. Nauka (in Russian).

Walker, M.W.;Kirschvinke, J.L.;Chang, S.R. and Dizon,A.E.1984. A candidate magnetic sense organ in the yellowfin tuna, *Thunnus albacares*. Science. 224:751-753.

Wallace, R.A. 1978. Oocyte growth in non-mammalian vertebrates. *In*:The Ovary. Edited by R.E. Jones. Plenum, New York. pp.469-502.

Warner, G.P. 1977. The biology of crabs, New York:Van Nostrand Reinhold.

Weiner, G.S.;Shreck, C.R. and Li, H.W. 1986. Effects of low pH on reproduction in rainbow trout. Trans.Amer.Fish.Soc.115:75-82.

Wiese, K.2002. Crustacean Experimental Systems in the Neurobiology. Springer, New York. 301pp.

Williams, M.C., 1977. Conversion of lamellar body membranes to tubular Myelin in alveoli or fetal rat lung. J. Cell Biol 72:260.

Xue. X.M., Anderson, A.J., Richardson, N.A., Xue, G.P. and Mather, P.B. 1999 characterisation of cellulose activity in the digestive system of the redclaw crayfish (*cherax quadricarinatus*). Aquaculture, 180:373-386.

Yamaoka, L.H. and Skinner, D.M. 1976. Free amino acid pools in muscle and haemolymph during the molt cycle of the land crab, *Gecarcinus lateralis* Comp. Biochem.Physiol. 55A:129-134.

Index

B

C

D

E

F

G

H

M

P

R

S

T

U

V

W

Y

Z

Zeitfracht Medien GmbH
Ferdinand-Jühlke-Straße 7
99095 Erfurt, Deutschland
produktsicherheit@kolibri360.de